Pre-Publication Review
Progress That Becam

"In times like these a ***new worldview*** c ... at the margins of power, at the periphery of the action unfolding on the main stage ... David Loye's central insight is in my opinion right on the money...The organizing principle of the new faith-a faith of human beings about human beings-is evolution itself. Not the traditionally taught evolutionary scenario dominated by competition and selfishness, but an understanding closer to the original Darwinian one that sees cooperation and transcendence of the self as ***the most exciting parts of the story.***"

> ***Mihaly Csikszentmihalyi,*** *co-founder of the field of Positive Psychology, Director Quality of Life Research Center, Claremont Graduate University, author of The Evolving Self: A Psychology for the Third Millennium, Flow: The Psychology of Optimal Experience, and Creativity: Flow and the Psychology of Discovery and Invention.*

❦ ❦ ❦

...***An amazing accomplishment!*** After *Darwin's Origin of Species* and *Descent of Man* left his hands, they reoriented our thinking about the world and the place of human beings in it. But only a part of the message of those volumes became emblematic of the Darwinian revolution. David Loye now focuses attention on the other part of Darwin's message, which is one of cooperation, brotherhood, and a progressive understanding of the still developing nature of human beings. His trilogy ... r***eveals what has been neglected with a passion and urgency not found in any other work.*** It is scholarship of deep humanity and needful wisdom, one that ***advances a new vision***, but one thoroughly in the Darwinian spirit."

> ***Robert J. Richards****, national award-winning science historian and Darwinian scholar; professor, University of Chicago; author of the classic Darwin and the Emergence of Evolutionary Theory of Mind and Behavior, The Meaning of Evolution, The Tragic Sense of Life, The Romantic Conception of Life, and Was Hitler a Darwinian.*

❦ ❦ ❦

"David Loye's is one of the few voices so desperately needed in the Darwin debates. Not only does he introduce "Eros" into the picture in a rational, sane, and supportable fashion, he ***makes the whole evolutionary***

theory hang together as "Eros-in-action," or, if you want, "Spirit-in-action" or "Love-in-action." It's been clear for quite some time now that the standard neoDarwinian synthesis can in no way account for the rise from dirt to Shakespeare, and new, believable theories are desperately needed ... The orthodox will cringe, and go on failing to explain evolution while bad-mouthing all other attempts, the Creative Intelligence folks will correctly spot all the missing holes in Darwinian theory and then-in an entirely unsupported move-fill in those holes by plugging them with Jahweh, a ridiculous move if ever there was one. The holes are supported by the data, Jahweh is not. David Loye's is both-spotting the holes, then filling them with something like a self-organization principle, which is a drive to higher levels of organization warranted by the data itself. Call that extra push "self-organization," or "Eros," or "Spirit," or "Love," or what you will, but it is fully justified by empirical data and scientific research. ***I know of no one doing this as thoroughly and carefully as David Loye. Read him, it's one of the most important topics alive today."***

> ***Ken Wilber,*** *pioneering philosopher, psychologist, and morally driven developer of Integral Theory; founder, The Integral Institute; author of The Atman Project, Integral Psychology, Integral Spirituality, Sex, Ecology and Spirituality, A Brief History of Everything, and The Integral Vision.*

❦❦❦

Darwin has been misrepresented and misunderstood: a fate not unique among great scientists and prophets. Their insights are made to serve their followers' aspiration and confirm their followers' worldview-never mind what the thinkers and prophets themselves truly had in mind. When David Loye goes after what Darwin had in mind he is not only putting right the historical record; he is ***performing a crucial service in the cause of humankind***. For today we, the species that calls itself homo sapiens sapiens, the knowing-knower, faces its most rigorous test of intelligence: the test of viability. Can we, will we, survive on our home planet? We don't know yet, but what we do know is that the answer lies in our ability to discern the path that lies ahead for evolution. ***David Loye lights that path for us, and for that we owe him a profound depth of gratitude*** -one we can best repay by comprehending what he has discovered, and acting on it. Doing so is in our most vital and immediate interest."

***Ervin Laszlo**, pioneering systems philosopher, scientist, and global activist; editor World Futures: The Journal of General Evolution; founder, the General Evolution Research Group and the Club of Budapest; author of Evolution: The General Theory, The Connectivity Hypothesis, You Can Change the World, and The Science of the Akashic Field.*

"Once in a decade or more a special book comes along, of urgent importance to the intellectual discourse of the time: Darwin, Freud, Jantsch, Lovelock. David Loye's *Darwin's Lost Theory* is this special. It represents the culmination of the Chaos Revolution, and the critical application of General Evolution Theory. It ***corrects an oversight in the history of science which has swerved the modern world off its track***. It provides the key to the reintegration of the sciences: physical, biological, and social. It can be the spark to jumpstart the social sciences to a new golden age of relevance to popular culture, by clearly showing how evolution theory bears on the survival of our species and our biosphere. I***n this work Loye has brought his unique erudition to an enormous and critical task, and carried it off with genius***. We urgently need this book, and we need it now."

***Ralph Abraham**, mathematician and chaos theorist, author of Chaos, Gaia, and Eros: A Chaos Pioneer Uncovers the Three Great Streams of History, Dynamics: The Geometry of Behavior, and The WEB Empowerment Book, Professor Emeritus, University of California at Santa Cruz.*

❦❦❦

"The idea that Charles Darwin himself believed that the final climb to human civilization required the enactment of a principle of moral conduct far above the "selfish gene" concept so prevalent in today's popular accounts comes as a surprise. But the fact that he argued at length and with passion for the recognition of this principle, along the way anticipating scientific concepts from far beyond his time, and further t***hat this work has been utterly disregarded by the official keepers of evolutionary theory ever since, boggles the mind.***

"Here, prominent social and evolutionary theorist David Loye treats us to ***a scientific mystery story of the first order.*** Taking us back to the final years of Darwin's life, in his home at Down and during the summer of 1868 at his Freshwater cottage on the Isle of Wight, where he struggled to find expression for the thoughts that would form the core of *The Descent of Man*, Loye leads us with sure steps through Darwin's emerging work, and through the Great Invisible Book that lies within, unfolding its vast implications and leaving no doubt that Darwin's long ignored plea for a larger vision of human nature is still relevant in the modern world and more desperately needed than ever.

"This is an immensely important book with an engaging and easy style that will recommend it to readers of all backgrounds and interests."

> ***Allan Combs****, psychologist and integral evolution theorist, author of The Radiance of Being and Consciousness Explained Better, professor, California Institute of Integral Studies. and Saybrook Graduate School and Research Center.*

❦❦❦

"In his book on Darwin's 'lost theory,' Loye ***grips the reader's imagination as if glued to watching him put together a giant jig-saw puzzle*** showing the whole sweep of evolution in the light of both former and recent thinking. I have been particularly fascinated by Loye's discovery of the connection between Darwin's projection of the evolutionary development of the moral sense and my own brain research. In the notebook of 1838 Darwin asked himself, 'May not moral sense arise from . . . our strong sexual, parental, and social instincts?' ***This is point for point what I found 100 years later in my own extensive exploration of the primate brain in regard to primal sex-related functions.*** I am very impressed with how Loye shows that Darwin expanded this core insight into the full theory so long overlooked in *The Descent of Man.*"

> ***Paul D. MacLean****, M.D., Senior Research Scientist, National Institute of Mental Health, evolutionary brain theorist widely considered one of the 20th century's two greatest brain scientists, author of The Triune Brain in Evolution.*

❦❦❦

"After a hundred years of evolution theory it's time to look at all of Darwin's writings-not just the ones encapsulated in the phrase "survival of the fittest." ***David Loye's account of contributions that revolutionize current evolutionary theory includes ignored ideas forwarded by Darwin himself***, now fleshed out in avenues not explored in Darwin's time. An example is ***complexity theory***, the finding of ***self organizing***

processes and how they apply to evolution by way of regulatory genes. All these new ways of approaching evolution do not as yet fit easily together, but the enterprise is young and Loye's contribution in calling attention to them will make it much easier to evaluate each and show how they can work together: a job Loye has heroically begun. This new way will shine light on our spiritual future, which rises out of a meaningful information revolution and leaves behind the now out-dated "material mankind" that was the heritage of the industrial revolution."

> ***Karl Pribram***, *widely considered one of the 20th century's two greatest brain scientists; developer of holographic and holonomic brain theory; former professor Stanford, Radford, and Georgetown Universities; author of Brain and Behavior, Languages of the Brain, The Form Within, and coauthor Plans and the Structure of Behavior*

❦ ❦ ❦

"If humankind is to survive the 21st century, it will need to deal far more effectively with the escalating threat of nuclear warfare, ecological devastation, and the other threats to its existence. There is no better roadmap ... David Loye's remarkable book ... has brought to light a rich vein of gold still to be mined in Darwin's writings. We're all indebted to his wisdom, his scholarship, and his careful articulation of Darwin's revolutionary paradigms, needed more now than ever before."

> ***Stanley Krippner,*** *pioneering humanistic psychologist; professor, Saybrook University; former president of the humanistic psychology and psychological hypnosis divisions, American Psychological Association; author of Human Possibilities and co-author of Healing States and The Realms of Healing.*

❦ ❦ ❦

"... ***absolutely brilliant***; one of the most important contributions in print about how we can re-understand human nature and re-invent a viable human society."

> ***Thom Hartmann***, *crusading journalist and host, national radio Thom Hartmann Program; author of Threshold: The Crisis of Western Culture, Unequal Protection: The Rise of Corporate Dominance and the Theft of Human Rights, and Rebooting the American Dream*

❦ ❦ ❦

"... has passionately called our attention to a part of Darwin's work that not only significantly modifies his construction of natural selection, but does so more prominently in *The Descent of Man* than many other modifications scattered throughout his vast writings. ***Even a number of neoDarwinians are now getting ready*** to accept some version of what Loye identifies as Darwin's discovery of 'organic choice,' usually under the label of 'self-organizing processes.' Loye's work comes along at a propitious time."

> ***Stanley Salthe***, *biologist and evolution theorist, author of Development and Evolution and Evolving Hierarchical Systems, Professor Emeritus, biology department, Brooklyn College of the City University of New York.*

❦❦❦

David Loye ***writes with passion of the more cooperative and empathic humanity that emerges in Darwin's long ignored development of a theory of the evolution of moral sensitivity*** ... also writes eloquently of the role that a language and philosophy based on the use of self-organizational concepts of systems theory can provide in pursuit of an evolutionary cultural trajectory to establish ... a global society. Among those working today to transcend simple reductionist evolutionary theory using complexity theory, he writes of his partner, cultural evolution theorist ***Riane Eisler,*** my brother ***Ralph Abraham*** and ***Stuart Kauffman***, the great evolutionary biophysicist.

> ***Frederick David Abraham***, *co-founder and past president, Society for Chaos Theory in Psychology and the Life Sciences; former research professor, UCLA, UC-Irvine, University of Vermont, and Silliman University, Phillipines; author A Visual Introduction to Dynamical Systems Theory for Psychology, and editor Chaos Theory in Psychology*

❦❦❦

"... an important contribution to illuminating the real bases of human social behavior. The complexity of our mental and emotional dynamical system ***argues against attempts to account for all human social customs and structures in terms of theories of the "selfish gene" or "sociobiology"*** variety ...Loye gets us into the heart of Darwin's works and shows that when it came to human evolution, love and connectedness were regarded not as anomalies but as intricately related to the entire evolutionary process. ***Sexuality*** has been assumed to be motivated solely by reproductive needs, and its pleasurable and bonding aspects discounted, whereas Loye shows that Darwin saw sexual evolution as the primary basis of bonding and love in many animal species including our own ... ***will fill an important gap. . I expect it will become one of the major books of the early Twenty-First Century***."

Daniel S. Levine*, theoretical psychologist and neural network theorist, author of Introduction to Neural and Cognitive Modeling and (forthcoming) Common Sense and Common Nonsense, professor, psychology department, University of Texas at Arlington.*

❦❦❦

"David Loye ***brings a brilliant career to its zenith*** ... With the world facing threats like climate change, Loye's profound insights into the evolutionary advantages of both competition and cooperation form twin pillars supporting the move to a sustainable global order. Long overdue, this new work completes Darwin's work."

William E. Halal*, leading futurist; professor, management science, George Washington University; president, TechCast LLC; author of Technology's Promise*

❦❦❦

"It seems to me David Loye is ***making a crucially important point in this book about politics and economics***. Here we are, presently caught in a fierce struggle between one kind of politics and economics aimed at advancing human evolution while another kind of politics and economics is hell bent on driving us backward. Yet going by all the evolution theory that science and education presently focuses on you would never know that this pivotal struggle ... to change things for the better in our world had anything to do with evolution. ***It looks like Loye ... will at last hammer this crucial point across.***"

John Robbins*, pioneering nutritional and social activist; recipient of the national Rachel Carson, Humanitarian, and Courage of Conscience Awards; author of Diet for a New America, The Food Revolution, and The New Good Life.*

❦❦❦

"Writing about love in a serious scientific context is a task reserved for thinkers who dwell on the sublime summits of human philosophy. By uncovering the long ignored emphasis on love in Darwin's original works, ***Loye reveals the deeply humane side to the evolutionary paradigm,*** which has been ethically degraded by the domination of concepts such as 'survival of the fittest' and 'the selfish gene.' Loye's writing is further ***a welcome combination of easy-flowing wording and solemn, old school elegance.***"

> ***Vuk Uskokovic***, *nanotechnologist and systems philosopher, University of California, San Francisco, and President, UCSF Postdoctoral Scholars Association*

❦❦❦

"... evinces an immense amount of research and apparent lifetime of passion for the deeper truths of evolution. Writing with erudition, ***Loye makes a convincing case that the moral sense is paramount in driving human evolution***."

> ***Steve McIntosh***, *Integral philosopher; author of Integral Consciousness, and Evolution's Purpose*

❦❦❦

"***David Loye is that rare scholar who investigates his subject beyond the normal parameters of science and history. His work on Darwin is brilliant and paradigm changing***. For more than one hundred years Darwin's work has been mainly used to lock in a superficial belief in 'survival of the fittest". In this new book Loye shows the depth of Darwin's true thinking and the importance of this new perspective for today's scholars and general readers."

> ***William Gladstone***, *author of The Twelve and producer of the film Tapping the Source*

❦❦❦

An astonishing work! Loye's passion for the essence of Darwin renders a great service ... ***brings us face to face with the very nature of being human-the caring, social, species-with-a-conscience mammal that evolution made us*** ...makes this ***a page turner*** that gives students, educators, scientists and policy makers a vivid history lesson ...***powered by love, insights and wisdom for our age and beyond*** ... shows us that in our battle for survival the regressive blinders can and must give way to the progressive urge in every human, in every culture. Our future depends on it.

Raffi Cavoukian,Singer, author, founder of the Centre for Child Honouring

❦❦❦

"David Loye ***has opened the minds and hearts of this whole generation*** to the real meaning of the world-changing work of Charles Darwin. Loye's work is vital to us all."

Barbara Marx Hubbard*, evolutionary visionary activist; President, Foundation for Conscious Evolution; author of Conscious Evolution and Emergence*

❦❦❦

"I read the whole book and love it! ***A must-read for all of us working for global transformation to a cleaner, greener, more equitable future for the human family***. This crowning achievement takes David Loye's important work in earlier books to a new level of synthesis, clarity, and power. It provides the missing history and context we need to align and bring coherence to many of today's movements for positive futures."

Hazel Henderson*, crusading moral economist, producer and host for the internationally distributed public television series Ethical Markets; author, Creating Alternative Futures, Building a Win-Win World, Beyond Globalization, and co-author, Planetary Citizenship.*

❦❦❦

"Loye's thesis is ***nothing less than revolutionary.*** In a carefully researched and beautifully written work, he dramatically changes our understanding of Darwin and of evolution itself."

Alfonso Montuori, *Chair of Graduate Studies, School for Consciousness and Transformation, California Institute of Integral Studies, Associate Editor, World Futures: The Journal of General Evolution, and author of Evolutionary Competence.*

❦❦❦

"Charles Darwin's Vision of Hope, ***Reborn***."

Kenneth Bausch, renaissance scholar; Executive Director, Institute for 21st Century Agoras; author of The Emerging Consensus in Social Systems Theory and Body Wisdom: Faith in Chaos

England

"This is the ***most exciting, most revealing book on Darwin that I have ever read.*** More than any other, it has restored the full grandeur to Darwin's thesis as it evolved, as living beings evolved, from the survival of the fittest, through altruistic acts in social communities to the final affirmation of a desire for good, more compelling even, than our desire for self-preservation."

Mae Wan Ho*, biophysicist and evolution theorist, author of The Worm and the Rainbow, Genetic Engineering, and editor, Beyond Neo-Darwinism: The New Evolutionary Paradigm, professor, biology department, The Open University, London*

Canada

"... ***a seminal step toward a critical pedagogy wedding bio-literacy and contemporary psycho-social scientific thought*** ...embraces controversy to radically expand basic Darwinian theory toward its yet unrealized true potential, as the action-oriented, functional scientific theory for the new millennium it was designed to be. This ***highly innovative interdisciplinary exploration of the span from biological to moral to spiritual evolution*** ...surprised me, sparked my interest, and changed the course for my scientific thinking ...bridging toward a transdisciplinary methodology that is our best, and perhaps only hope of successfully sorting through future best-fit models.

Christopher Peter Montoya, *professor, psychology department, Thompson Rivers University, Canada; author of The Migratory Theory of Genetic Fitness*

Finland

Many scientists still regard NeoDarwinian evolution theory as the complete scientific account of the origin and future for human life. But through carefully analyzing what Darwin actually wrote, and its corroboration by hundreds of modern scientists, David Loye ***finds shocking evidence of a higher order, open ended evolution theory seeking to transcend and replace "survival of the fittest" Darwinism***. Moral sensitivity far outweighs selfishness, mutuality is more meaningful than competition, love outpaces survival of the fittest a hundredfold .... ***we're driven to move*** beyond a science committed solely to the passive role of the so-called objective observer ***to the active role of science as partisan on behalf of and advocate of humanity*** ... what a full spectrum, action-oriented theory of evolution should look like, and how to actually build it.

> ***Pentti Malaska***, *pioneering futurist; professor of mathematics, Turku School of Economics, Turku, Finland; founder, Futures Research Center, and chairman, Academy for Futures Study Network*

Chile

"One of the central difficulties in modern biology is how to account for the origin of those human features we are inclined to consider superior, traits such as morality, ethics, rationality, self-consciousness, and spiritual experiences. The difficulty is that they must have arisen in evolution from a manner of living that did not contain them ... shows that Darwin saw this, and that his vision of a detailed answer to the question in terms of human emotional and cognitive development beyond the basic operation of natural selection has not been acknowledged. It is i***mportant that this part of Darwin's writing be recovered, as Loye does very clearly and in a compelling manner in Darwin's Lost Theory***."

> ***Humberto R. Maturana***, *professor, Department of Biology, The University of Chile, developer of the concept and theory of autopoiesis, author (with Francisco Varela) of Autopoiesis and Cognition and The Tree of Knowledge, and (with G. Verden-Zoller) of Amore e Gioco and other books in Italian, German, and Spanish.*

Taiwan and Australia

"David Loye, as expected, does a brilliant job of deconstruction and reconstruction. He takes the conventional view of Darwin, and much like Hegel did of Marx, turns him upside down. We learn that it is not survival of the fittest that is the thrust of Darwin's work but moral sensibility. This re-reading of Darwin provides us with a theoretical framework for a new biopolitics and a new vision ***of the future***. Written in an easy to read style, Darwin's Second Revolution is recommended for the scholar as well as the day to day facebook web surfer. Brilliant. And fun."

> ***Sohail Inayatullah***, *pioneering futurist; professor, Tamkang University, Taiwan and the University of the Sunshine Coast, Australia; editor, Youth Futures: Empirical Research and Transformative Visions, Macrohistory and Macrohistorians, and Chaos and Coherence in our Uncommon Future*

❦❦❦

New Zealand

"... ***a major scientific treasure*** ...Carefully piecing together fragments scattered in The Descent of Man ... Loye reconstructs the 'hidden' theory and shows that ***Darwin believed that love, rather than the "selfish gene," is the prime mover in human evolution*** ... should cause a revolution in social theory as diverse fields such as human ecology, urban studies, population dynamics, collective organization, and the study of culture and moral order are rethought and recast in the light of Darwin's moral theory. ***Darwin's Lost Theory*** is absolutely ***essential reading for every social scientist***."

> ***Raymond Trevor Bradley***, *sociologist, Director, Institute for Whole Social Science, New Zealand, Associate Research Professor, BRAINS Center, Radford University, Radford, VA, author Charisma and Social Structure: A Study of Love and Power, Wholeness and Transformation.*

❦❦❦

China

"At the funeral of Karl Marx, Friedrich Engels said that, as Darwin had discovered the evolutional rules in biology, Karl Marx discovered the evolutional rules in human society. Since then human social practice has falsified Marx's revolutionary theory and human scientific practice has enriched Darwin's evolutionary theory. David Loye's work ***exploring the relation of social and cultural evolution as well as biological evolution to human practice*** over the past two centuries is ***a meaningful advance for Darwin's theory***."

> ***Min Jiayin****, systems philosopher and research fellow, Institute of Philosophy, Chinese Academy of Social Sciences, Beijing; author of Evolutionary Pluralism: A New System of Systems Philosophy, and editor, The Chalice and the Blade in Chinese Culture*

China

"... ***blows the lid off the Darwinian delusion that Darwin was as blindly mechanistic as some of his neo-Darwinian successors in evolution theory.*** This new story of Darwin comes at a time when the dominant narrative of mechanistic biology has hit the brick wall as an explanatory theory ...Through carefully researched case studies, David Loye shows that Darwin was a genuinely humane, even spiritual human being ... ***shows that the dry rationality of contemporary mainstream science and academia have distorted the truth of one of the great men of history***. Modern science depicts Darwin as a stubbornly courageous purveyor of truth, struggling to light the darkness of religious and social superstition ... Loye's account of the real Darwin is a call for a deeper enlightenment, ***a powerful ray of hope***."

> ***Marcus T. Anthony****, founder, Mind Futures (www.mindfutures.com), Hong Kong; author of Integrated Intelligence, Sage of Synchronicity,and Extraordinary Mind: Integrated Intelligence and the Future*

Rediscovering
DARWIN

BY THE AUTHOR:

The Healing of a Nation

The Leadership Passion

The Knowable Future

The Sphinx and the Rainbow

Arrow Through Chaos

The Evolutionary Outrider, Editor

The Great Adventure, Editor

Measuring Evolution

Darwin's Lost Theory

Darwin's Second Revolution

Darwin in Love

3,000 Years of Love

100 Days of Love

1001 Days of Love

Brave Laughter

Return to Amalfi

The Partnership Way
(with Riane Eisler)

Rediscovering DARWIN

The Rest of Darwin's Theory and Why We Need It Today

DAVID LOYE

ROMANES PRESS
2018

PUBLISHED BY ROMANES PRESS

Print ISBN: 978-0-692-98402-4
eBook ISBN: 978-0-692-06064-3

1. Science. 2. Evolution. 3. Religion. 4. History.
5. Philosophy

Cover by John Mason

Print and eBook production
by David Gordon / Lucky Valley Press
www.luckyvalleypress.com

For more information:
Romanes Press
P.O. Box 51396
Pacific Grove CA 93950
phone: 831-626-1004
email: elliotsanders@gmail.com

Printed in the United States of America on acid-free paper that meets the Sustainable Forestry Initiative® Chain-of-Custody Standards. www.sfiprogram.org

A NOTE TO THE READER*

(i.e., progressive scientists, science writers, reviewers, historians, philosophers, theologians, ministers of all faiths, newsmen and newswomen, general readers, and oncoming generations)*

This book is being rushed into print because here, in page after page of his own words, you'll find Darwin's powerful refutation of the survival of the fittest/selfish genes/and now winner versus loser mindset driving our species and planet toward destruction --- and what to do about it.

Written by a scientist to gain a general as well as scientific readership (see above), this advance publication of Rediscovering Darwin is stripped down to the gripping story of the discovery of the long buried higher order rest of Darwin's theory of evolution.

Reconstructing Darwin will add the notes, references, index, for the wedding of both halves of Darwin's theory hailed by the remarkable body of scientists, thinkers and activists whose reviews open this book.

Revolutionizing Darwin will report the results of the ongoing development and testing of a new Darwinian way of measuring and potentially accelerating our species' evolution.

To Riane, as always

CONTENTS

PROLOGUE

DARWIN AND THE WORLD THAT COULD HAVE BEEN

Why take a new look at Darwin?

After 100 years of thousands of books in at least 100 languages, poked through by five generations of scientists and other scholars, could anything still be left of relevance to us today?

To begin, what is "survival" and who are the "fittest"? What's driving the new rise of the old politics of hate, the economics of all-for-me, the ice melting, seas rising, monstrous new hurricanes, the rampage of the new authoritarians, and the round the clock threat of nuclear annihilation, and the rest that drives the body blow to sanity that shoves this question at us?

Where did we go off track, and how can we get back on track in evolution?

Here for the first time at length, in page after page, is what Darwin originally wrote about his great theory of evolution.

But this time we'll not stop with only the magnificent first half of his theory. That was established with publication of *The Origin of Species* in 1861. Ten years later, hidden within Darwin's *The Descent of Man*, came the completion of his grand vision of *human* evolution—and then the incredible burial by the mindset of "survival of the fittest" and "selfish genes" still everywhere devastating our lives and time.

What follows is the story of the discovery of the hopeful higher half of his theory that Darwin scattered throughout the sprawl of *The Descent of Man* and the mystery of how on earth it was so effectively erased for over 100 years.

How could it go for so long unnoticed that in Origin's classic sequel on *human* evolution, Darwin wrote only twice of "survival of the fittest" but *95 times about love.* And this with only a single trivial entry for love in the loveless index still in *use worldwide.*

Even more remarkable in a battered world desperately in need of moral guidance is the Darwin who wrote 92 times not of selfishness but of the Moral Sense as the ultimate over-riding prime driver of evolution.

Yet here we are, caught within the grip of the Darwin sold to us as scientific proof of the cynical claim that the worldwide moral imperative of "do unto others as ye would have them do unto you"—that is, caring and concern for the welfare of others, A.K.A. *altruism*—is nothing more than a wily artifact driven solely by selfishness and selfish genes.

Why has it taken so long for us to see, let alone solve, what for so long has in effect held us captive within an immense scientific and social murder mystery?

Darwin and the World That Could Be

Who did it? How? Why?

I've written this book to reach both general readers and my fellow scientists. Despite the difficulties of reaching both in a single book, I've done this because it will take the drive of a rare new level of our mutual understanding if we are to push on past all that in blindness and madness is trying to drive us backward in evolution.

How can we speed up the evolution of our species and slow down the devastation of our planet before it's too late?

It seems impossible, but imprisoned within the denial of what Darwin most deeply valued is the crippled giant who still lives in our lives. And here, in his own words, out of chapter after chapter recovering the rest of his theory, the "lost"rest of Darwin's theory unfolds.

In ***Part I: A Startling Discovery*** you'll find the foreshadowing of what became today's immeasurably greater threat of nuclear annihilation. How during the Cold War scientists from both sides met secretly in Budapest to see if we could replace the survival of the fittest mindset, then driving us toward destruction, with a better theory of evolution.

You'll see how we set out to gain order out of chaos with chaos theory for a starter. And the buried history spanning the 19th, 20th, into the 21st century, which shows how the rest of Darwin was, with the best of intentions, wiped off the slate of history.

Part II: Recovering the Rest of Darwin's Theory explores the past, present and future of all that his well-intentioned but tunnel-visioned successors dropped from Darwin. A new understanding of ***sex, love, community,*** and of greatest importance, the eternal prime guidance of the ***moral sense***.

Here you'll also find the shock of Darwin's long ignored case for ***spirituality*** and the place and function of ***religion*** in evolution. And the surprise of how, in what he wrote of "the morality of women," Darwin even became a cautious forerunner of male support for the women's movement.

It reads like the roll call for all that really matters!

You'll also find what happens if you set out to find out what buried the rest of Darwin. How this book that begins with the overtones of a spy novel, becomes a detective story, becomes a murder mystery, and ultimately a scientific story of adventure and romance.

In short, you'll find the buried science and story of how we lost the chance to gain the *world that could have been*. But instead got *the world that is.* But now still have a fighting chance to gain *the world that should have been* by joining the drive of all those among us—well known by name and works—who serve as earth's new explorers of the better world, who you'll meet, and get to know in this book.

And this too, of greatest meaning to me personally. In completing this book in my nineties, in looking ahead and resonating to the wonder and horror of life over much of the 20th into the 21st century, it's the course I hope this book will take into the future.

We older ones are too locked into our own agendas to get to where this book shows we not only need to but must go. But among our students, and their students, and on down the line for significant others coming new to this task, are those who in this book can find what they need to push on and get there.

It is our job to identify them, and encourage them to make this cause theirs, and to support them in every way we can. But it will take the fresh, eager, indomitable arousal of oncoming generations, in joy riding on the vision that drives our species, to get us back on track in evolution.

You will know who you are when the time comes.

Bon voyage!

PART I

A STARTLING DISCOVERY

ONE
THE ENTRY POINT

Uncovering what happened—that is, how, why, and who was involved in burying the rest of Darwin—has not been an easy task. It took a threat of nuclear annihilation, a secret meeting of concerned scientists, formation of our multinational advanced research group, and then the shock of a sleepless night and twenty going on now thirty intensive years of further investigation.

The entry point for what became our somber but increasingly buoyant adventure arrived after a very short stretch of hope for peace after WW II.

For me it began with a phone call to my research office at the UCLA School of Medicine. It was a deep voice with a thick Hungarian accent inviting me to fly to Budapest all expenses paid for a secret meeting of concerned scientists from both sides behind the Iron Curtain.

As we flew in from London, nearing Budapest, my partner and I could see way down there below us the barbed wire fence that, extending out from the infamous Berlin wall, literally sliced Europe in half.

We moved past grim Russian guards with machine guns and the fanged welcome of snarling attack dogs on leash. We spent the night in a dismal Soviet era structure that looked like, and within felt like, an old time ice box. Next morning, within the dingy, grey-yellow crumbling mass of what had once been the majestic Hungarian parliament building, we met the others summoned from both sides.

It was a bare walled room in what seemed to be the basement, I recall, with a single bare lightbulb overhead trying, as if in apology, to cut the gloom.

On one side of the table sat the handful of us who had flown in from the West. Across from us sat the handful from the East—including a big, black-bearded bear of a fellow who turned out to be the Director of Russia's most prestigious research center. Alongside him were what turned out to be two captive Hungarian scientists.

We sat there for a while in silence waiting for the mysterious figure who had called us together to arrive and introduce us.

The minutes ticked away. We sat there uncomfortably, occasionally staring at one another. Then out of the shadow beyond the reach of the wan light bulb, as if with the flash of a cloak in a Dracula movie, he was there before us. Lean, wiry, with elfin face and a high forehead flanked with white wisps, his dark deep-set eyes scanning to grab the eyes of the rest of us—this was Ervin Laszlo, considered the world's foremost general evolution theorist.

In science chaos theory was just beginning to take hold. With mesmerizing eyes Laszlo laid out before us a daring plan. The idea was we'd use the new power and popularity of chaos theory to see if it could effect a crucial update for evolution theory.

We would band together to see if we could come up with something that could end the reign of the bloody, ostensibly Darwinian "survival of the fittest" mindset, then driving Russia and the United States and all the rest of us toward destruction.

Quietly, at first secretly and then more openly drawn together from both sides of the Iron Curtain, our group came to include scientists from both natural (i.e., physics and biology) and social science (e.g., psychology, sociology, economics and political science), as well as history and philosophy.

From that first gathering of a dozen of us—from Russia and Hungary on their side, from England, Finland, and the U.S. from our side—came the formation of our General Evolution Research Group. Under Laszlo's canny leadership, GERG (which out of much merriment became our playful acronym) soon expanded to include prestigious members from throughout the rest of Europe clear on out to China.

For nearly a decade we had labored on this magnificent mission, but seemed to be getting nowhere. Many nights, tossing about, unable to sleep, I was hounded by the feeling that something very big was missing in the great tangled mess of theory we were wrestling with.

I thought of Charles Darwin—and the irony of how here was this great man, the revered founder to whom we all paid lip service, but whose books few of us bothered to actually read any more. By our slam bang modern expectations *Origin of Species* and *The Descent of Man* were cumbersome, wordy, quaint, sure to be old hat stuff for which none of us had the time. So literally almost all of us—not only in our group but everywhere—relied on secondary sources.

That is, our thoughts, and discussions, and writing, and teaching was based on what others told us was Darwin, the whole truth and nothing but the truth. But only in a lonesome quote here and there were we given what he himself had said.

Could there be something out there in Darwin we were missing?

TWO
DISCOVERY!

I had a computerized copy of *The Descent of Man* in which Darwin specifically tells us this:

> In consequence of the views now adopted by most naturalists, and which will ultimately, as in every other case, be followed by others who are not scientific, I have been led to put together my notes, so as to see how far the general conclusions arrived at in my former works were applicable to man.

Clear enough. He was moving on from prehuman evolution in *Origin of Species* to the world of that hypothetically more advanced creature—*homo sapiens sapiens,* our selves. But what else might lie on beyond in the rest of Darwin?

Once again I couldn't sleep. I lay there hounded by the feeling that while the world was falling apart all around us, not only within our own group but everywhere, science was only nibbling at the task of building an adequate evolution theory.

A theory not just for theorists, but one that could provide us with a practical guide to the future for everybody.

Could there still be anything in Darwin bearing on this increasingly crucial task?

Again I told myself I would look into it in the morning. But still I couldn't sleep.

At last I got up, put the disk in, clicked to *The Descent of Man,* then hit Search to see what Darwin had to say about the first thing that came to mind.

What should it be?

Survival of the fittest, of course.

What did Darwin have to say about what over 100 years has worldwide become, for millions of us, the quick answer—and *complete* answer—for what drives us in Darwin's great theory of evolution.

And there on the screen before me it was— the fact that in 800 pages of fine print, in *The Decent of Man*, of "survival of the fittest" Darwin wrote only *twice,* once to regret ever using the damned phrase.

In shock, I tried for what seemed the most fitting of likely polar opposites. And I found he'd written *95 times about love*!

But what about the "selfish gene" and the best-selling books with the booming new focus on selfishness that were adding a sharp new cutting edge to "survival of the fittest?"

Of selfishness I found that Darwin had written only six times, and in a veritable roar out of a page in *Descent,* he called selfishness "a base principle" accounting for "the low morality of savages."

What might be the polar opposite to selfishness? Most logical would be caring for others, or the driver of moral behavior we call *altruism*. And of everything to do with the word moral—i.e., moral sensitivity, moral belief, the moral evolution of ours and preceding species— *I found Darwin had written 92 times!*

Twice of "survival of the fittest" but 95 times of love. Six times of selfishness but 92 times of moral sensitivity.

What did this tell us?

Of *competition* I found he'd written 12 times, but of *cooperation* (called mutual aid in his time) over twice as many with 27 times.

Here are the word counts I found for more of the powers of mind that build everything we know, delight in, and cherish as civilization.

"Mind," 90 times
"Intellectual qualities and powers" = 58
"Intellectual powers" = 17
Reason = 53
Imagination = 25
Learning = 18
Consciousness = 15
Curiosity = 14
Instruction = 10
Brain = 110
Habit = 108

What happened? How did so much get not merely shelved but actively buried?

Considering all that has worked out not as we dreamed and fought for, but rather as the unfolding nightmare of all too much of our time, what are we looking at?

Where, why, how, and *who* did it?

THREE
SELF-ORGANIZING

Where to begin?

Here was this great sprawling book, which Darwin tells us is on *human* evolution, *The Descent of Man*. Few bothered to read it any more, but I found it riddled with clues indicating something very big had been tucked away out of sight.

It did indeed look like what one might rightly call a social scientific murder mystery.

At the time of the Budapest meeting, and formation of GERG, the leading edge in science was the acclaimed "new" territory of *self-organizing processes*.

The great bit of luck for us was that Laszlo had gained four of the major figures in the development of self-organizing process theory for our group.

The Nobel prize winning Belgian thermodynamicist Ilya Prigogine had shown how the evolutionary thrust of self-organizing works at the molecular level through the operation of what he called *autocatalysis*.

With eminent Chilean biologist Humberto Maturana, Francisco Varela—Sorbonne based in Paris during the years of his involvement—had sharpened the picture for self-organizing in the biology of perception and cognition with their concept of *autopoiesis*.

And the two Russian captives who were there, and had joined in our founding, eminent Hungarian biologist Vilmos Csanyi and his student and partner, computer scientist Gyorgy Kampis, had moved self-organizing on up into systems science and cultural evolution with the wide-ranging implications of what the

Hungarian duo called *autogenesis*.

Further with us, decisive in the development of self-organizing theory, was Karl Pribram, the great crusading brain scientist who uncovered the pivotal understanding of the *active* versus reactive brain and mind.

Later, further expanding the reach of our group, was the noted mathematician and leading chaos, complexity, and self-organizing theorist Ralph Abraham.

They were, for our time, the stars, the Supermen who had shown how evolution was not just a matter of our body, brain and mind being chosen—stamped out, or otherwise exclusively shaped by something *external* to us. Counter to the prevailing mindset for evolution they had shown why and how we were not just the passive product of natural selection, "blind chance," and "selfish genes."

They had uncovered the multi-level drive within us that opened not just to our species, but in varying degrees to all species, the choice and voice in shaping who we were, and are, and what our future can be.

Could it be that something like their insight had been lost to us in what increasingly looked like the burial of a vital chunk of Darwin's original theory of evolution?

On intensively digging into what Darwin had written in *The Descent of Man,* there it was. Yes, indeed. Over 100 years ago Darwin had accurately perceived and even re- emphasized what had become the celebrated discovery of advanced science for our time.

Here, buried in plain sight—set off here within a text box, as will be the case for all quotes for Darwin—was his pioneering perception of what in our time became self organizing process theory.

Darwin on Self Organizing Processes

Besides the variations which can be grouped with more or less probability under the foregoing heads, there is a large class of variations which may be provisionally called spontaneous, for to our ignorance they appear to arise without any exciting cause.

It can, however, be shewn that such variations, whether consisting of slight individual differences, or of strongly-marked and abrupt deviations of structure, *depend much more on the constitution of the organism than on the nature of the conditions to which it has been subjected. (ital added)*

Mr. Wallace, in an admirable paper before referred to, argues that man, after he had partially acquired those intellectual and moral faculties which distinguish him from the lower animals, would have been but little liable to bodily modifications through natural selection or any other means. *For man is enabled through his mental faculties "to keep with an unchanged body in harmony with the changing universe."* (ital added)

He has great power of adapting his habits to new conditions of life. He invents weapons, tools, and various stratagems to procure food and to defend himself.

When he migrates into a colder climate he uses

clothes, builds sheds, and makes fires; and by the aid of fire cooks food otherwise indigestible.

He aids his fellow-men in many ways, and anticipates future events.

Even at a remote period he practised some division of labour.

The lower animals, on the other hand, must have their bodily structure modified in order to survive under greatly changed conditions. They must be rendered stronger, or acquire more effective teeth or claws, for defence against new enemies; or they must be reduced in size, so as to escape detection and danger.

When they migrate into a colder climate, they must become clothed with thicker fur, or have their constitutions altered.

If they fail to be thus modified, they will cease to exist.

Like the rise of the Genie from the magic lamp to Aladdin's bidding, this finding affirmed the value of further pursuit in this direction. It lit up much more of what lay behind the clues of that sleepless night.

I began to see how after the death of Darwin evolution theory took off in two fateful directions—and that, in turn, began to reveal how, why, and who buried the rest of Darwin.

There was the great saga of the evolution of our and all prior species told through development of Neo-Darwinian theory by

Darwin's direct line successors in biology. This triumph thousands of books have explained at length and rightfully cœlebrated.

The first half of his theory, the *Origin of Species* part, was in reasonably good shape before corruption into the rampage of the survival of the fittest/selfish genes mindset. But unsettling, and to me increasingly horrifying, was the degree to which we are now living in the disastrous down side of all that was lost in the burial of the rest of Darwin's theory.

In science we tend to feel that something isn't really real unless it can be pinned down in a flurry of abstractions, but in the case of self-organizing process we're seeing the earlier power of Darwin's simple concrete descriptions.

Being able to describe, probe and put to use what now seems the obvious fact of self-organizing process with the incredible power of mathematics has been a great gain for science. But after I gained an understanding of the work of Prigogine, Varela, Csanyi, Pribram, and Abraham, what struck me was how simple, at its core, the idea is. Beneath the complexity of all the mathematics and differences in methodologies and languages used to define self organizing was what the term itself so plainly tells us.

It is this matter of us humans having a voice in the process. It is the boost of seeing something far beyond biology's shuffling of genes in variation.

What we're looking at is the vital matter of the power within us of *choice* in the shaping of who we were, are, and could and should become that's missing in the current thrice-blessed theory of evolution we're caught in.

In fairness you could argue that one could hardly expect Darwin's successors to bypass nearly 100 years of the development of science to catch up to where research groups like ours had finally arrived. But if we put to use the new evolutionary systems science—the bond in common for our group—much out beyond the boundaries of traditional biology comes to light.

We find how, under a variety of names, the evolutionary function of self-organizing process has century after century been obvious to philosophers, educators, novelists, and—simply in our everyday experience—to most of us.

Self-Organizing Writ Large

I found that once you're pointed toward this pioneering perception, many more answers to the mystery of what happened to the rest of Darwin open out of insight piled on insight.

I came to see how behind the term lay an exploration going back over 2000 years to the Greek philosophers —in Pythagoras and Empedocles, for example.

I saw how for well over 100 years, under a variety of other names, self-organizing process had actually been the driving force behind the development of the fields of psychology, sociology, economics, political science, on out through all the rest of the social sciences.

It had been the same for teachers and educators, from Johan Pestalozzi to Maria Montessori, John Dewey, and others of our time.

Indeed, within the super-sensitive probe of the humanities—which so often in poems and novels range far ahead of science—are celebrations of self-organizing process.

There are, for example, the opening lines of the famous poem Invictus, by the "great, glowing, massive-shouldered fellow with a big red beard, with a laugh that rolled like music," British poet W.E.Henley.

Out of the night that covers me
Black as a pit from pole to pole
I thank whatever gods may be
For my unconquerable soul.

Going back within the 100,000 to 200,000 year experience of our species, wasn't this capacity also what nurtured, buoyed and bucked us up to go on?

It's difficult for those of us disciplined by science to display anything smacking of even the least bit of enthusiasm or, perish the thought, passion in any form. But if we cast aside the academic armor for a moment, do we not also see here the driver for Jesus, Gautama, Mahatma Gandhi, Martin Luther King, Elizabeth Cady Stanton, Dorothy Day, Sojourner Truth, and on and on with everyone who fought against tyranny, ignorance, and oppression?

Memories of the Golden Years of GERG

And what did the members of the General Evolution Research group we've met so far look like?

A touch of description here and there may violate the custom for scientific writing, but if this will bring what I write to life, so be it. Let's see ....

To my surprise Prigogine was not at all what I expected. Instead of the stereotypical bespectacled and Einsteinian wild white-haired man of science, his hair was thin, black and slicked back, and he looked more like a portly Belgian CEO addressing the Board of Directors for a monster corporation.

Varela was a neat, trim, handsome fellow who, though brilliant, I found awfully annoying. For he would jump in to disagree every time we seemed to be nearing a consensus on anything—rather essential for our or any other project to succeed.

Karl Pribram was a romantic figure with flaring white hair, bold assertive face, who, with the rapier like thrust of corroborating agreement or disagreement, looked like he could have been one of Dumas' three musketeers.

Csanyi had a voice that seemed to roll around his tongue on the way out. With his heavy black beard and wild shock of clipped black hair, he looked like the stereotypical bushy bearded anarchist.

After that first meeting Czanyi took us carousing, with the Russian bear and the merry Finn Penntti Malaska, through the wonders of Budapest at night.

We dined in the restaurant with the plink and twang of the *bousuki,* the open-faced piano made famous by Orson Wells in the movie The Third Man.

As if on cue, the gypsy violinist swooped in to further transport us into the dark moody mystery of the gangster Harry Lime. And on ahead we were startled to hear banjo, guitar and fiddle. There in the Gothic cobbled center of Budapest, there in the moonlight, we found a band of Hungarians were playing indisputably authentic American blue grass music.

Last then Csanyi took us to his home to see his pet fish—a supposedly million or so year old giant mournfully bug-eyed fish he kept in a giant tank—and to meet his wife, a lovely violin player in the Budapest symphony orchestra.

I mean by this expression that the whole organisation is so tied together ... that when slight variations in any one part occur, and are accumulated through natural selection, other parts become modified.

Darwin, *The Descent of Man*

FOUR
CHAOS THEORY

If self-organizing theory was foreshadowed in the buried rest of Darwin, what about chaos theory?

A major tenet of chaos theory is "the butterfly effect"—first described in 1972 by the eminent mathematician Edward Lorenz in his famous paper "Predictability: Does the Flap of a Butterfly's Wings in Brazil Set Off a Tornado in Texas?" This vivid mathematical discovery soon became useful in many situations where a slight disturbance in one place in a system can lead from slight to enormous consequences elsewhere.

Possibilities focused on the impact of the "butterfly effect" on large weather systems, then later within economic, political and social systems. Could the link between loss of the rest of Darwin and the threat of nuclear annihilation also be here?

Yes, indeed—to which I would add an exclamation point. For there again was Darwin's probe of the same force at work.

Darwin on Chaos Theory

The law of correlated variation, the importance of which should never be overlooked, will ensure some differences; but, as a general rule, it cannot be doubted that the continued selection of slight variations, either in the leaves, the flowers, or the fruit, will produce races differing from each other chiefly in these characters. I mean by this expression that the whole organisation is so tied together, during its growth and development, that when slight variations in any one part occur, and are accumulated through natural selection, other parts become modified.

This is a very important subject, most imperfectly understood, and no doubt wholly different classes of facts may be here easily confounded together.

This is how Darwin first observed in *Origin of Species* the "butterfly effect" of chaos theory at work in what he called "correlated variation."

To try to make sure his successors get this idea, Darwin took another whack at it.

I will here only allude to what may be called correlated variation ... long limbs are almost always accompanied by an elongated head ... cats which are entirely white and have blue eyes are generally deaf.

It might be a crazy notion, I thought afterward. But by some stretch of the imagination, could we use this weird link between Darwin and chaos theory to head off nuclear annihilation for our poor species?

I puzzled over it for while, then gave up. But where now?

Living in the deep woods north near Santa Cruz was the University of UC Santa Cruz professor and purported wizard of chaos, Ralph Abraham.

It was only an hour away from my home in tiny Carmel-by-the-Sea. While on an international speaking tour Laszlo had dropped by to see us and catch up on GERG doings. So Laszlo, my partner Riane Eisler (whom we'll shortly meet), and I went to see Abraham and soon enlisted a new member for our fledgling General Evolution Research Group.

Soft spoken, precise in words and movement, with a beaming face wreathed in a halo of white hair and beard, behind Abraham lay the thousands of years since Pythagorus to advance the power of measurement that mathematics brought to science.

As we moved from the Butterfly Effect into what seemed to be chaos theory's powerful new way of understanding a good bit of everything in our lives, Ralph's involvement became invaluable.

Also defined by Edward Lorenz in his pioneering analysis of weather systems—further affirmed by centuries of impeccable mathematics, and blessed with the romance of its founding in the exotic work of the French whiz Henri Poincare—I found that chaos theory was a matter of *attractors* that pull us in three directions.

First is the *static attractor,* which reaches out to pull in everything together within a system to keep things fixed in place.

You can see this operating almost any day in the sky, where out of the moisture the wisps are pulled in to form a cloud.

Next comes the *periodic attractor*. Working like the pendulum that swings right to left, left to right, back and forth in an old clock, here we see the see-saw familiar to us in politics and other areas of our lives.

Documented in history by Arthur Schlessinger, Sr., as well as in my own work in the psychology of ideology, here was the force at work in the shift from liberal to conservative eras, conservative to liberal eras, back and forth, over and over again through time.

Then comes the *strange or chaotic attractor,* where in effect everything busts loose.

Clouds become hurricanes. Assassinations become wars. Titanics sink and terrorists topple the World Trade Center. But also Copernicus maps a new planetary arrangement. Jesus is born. The French Enlightenment and the American Revolution challenge the "divine right of kings." New paradigms burst out of old paradigms, and we get everything else in the arts and in life that winds up being classified as good or bad for us.

FIVE
ORDER OUT OF CHAOS

And yet, as so often happens, there was more here that I should add.

For it was out of this much larger mind space that I first glimpsed how the bloody history of our species could have been changed for the better had the rest of Darwin not been ignored.

I saw how, by in effect making evolution theory their exclusive property well into the 20th century, biology and paleontology had in effect scuttled the alternatives.

They had powerfully advanced our understanding of the past and biological evolution. They had put this new understanding effectively to work in agriculture, medicine, and other areas advancing the health and well-being of our species and our planet. But in orienting to the past, they had shut out the future.

While on one hand they were advancing our species, on the other was the tragic irony of how, blindly and unintentionally, they were diminishing our species.

The quickest way I can make this vital point is through this comparison. All around us we can see how, despite the amazing exploits of *natural science,* the *theory* of evolution only very slowly creeps ahead.

But within *social science* the *real world* understanding of evolution has *raced* ahead.

But almost none of this vast warehouse of striking findings is being picked up and put to use in what is identified, taught, hailed, and ostensibly advanced as the official one and only evolution theory.

Freud, Jung, James, Lewin, Maslow in psychology. Weber, Durkheim, Merton, Sorokin in sociology. Boas, Benedict, Mead, Montagu in anthropology. These and countless other founding social scientists explored and wrote of the real life drive of love, the moral sense, and other factors that Darwin identified as crucial in human evolution. But all failed to make the cut for what was universally taught and cited as the one and only evolution theory.

Here, one could say, was the second murder and burial.

That is, as if right on schedule for the standard murder mystery, here was a trans-historical replay of what earlier happened to the higher order rest of Darwin.

The first victim of had been Darwin himself. Now everything with an investment in the politics and economics of the survival of the fittest/selfish genes/winner versus loser mindset that was destroying us and our one and only planet, conspired to end the job.

Artful demagogues rallied the crushing weight of the moneyed few to pack on into the grave anything that might in any possible way threaten their drive for proper worldwide rule. Wild as this may sound to anyone who wasn't there and involved, this was roughly the situation until out of a wave of discordant insights in many fields burst the chaotic attractor, which in mid to late 20th century collapsed a world of dangling meanings into the brief, wide, surface popularity of cybernetic, chaos, complexity, and self-organizing theory.

In effect, this was the broom that for a precious time swept clean and opened the mind of science and society to the long delayed but vital next step for the development of an adequate theory of evolution.

It was the insight that spurred me to take the step on beyond what knocked him in the head, then buried the rest of Darwin, which still lay hidden behind the clues.

My Own Contribution

It seemed self-puffery if I left it in the last chapter, but needlessly self-effacing if I left it out, so I decided to put this personal bit here in a box in neutral territory.

As a psychologist with a sub-speciality in sociology within a research group primarily composed of biologists, physicists, and mathematicians, I was struck by what happened when you crossed over from natural to social science.

Out of the new multi-level understanding of evolutionary systems science, which transcends differences and bound us together within our group, I suddenly saw how the "strange attractor" of chaos theory, then embedded in biology and mathematics, was mirrored in the awesome, mysterious and imperative force that drives everything from great scientists and leaders to movie stars.

Driver of creativity, herald of evolution, I saw what the great co-founder of sociology, Max Weber, described and called "charisma."

I saw how the "self organizing processes" of mathematics and biology crossed over from natural science into social science to become the "functional autonomy" of the lovable psychologist Gordon Allport, as well as the charisma explored by a member of our group, New Zealand sociologist Raymond Bradley.

In what became a pivotal paper in the influential journal *Behavioral Science,* in our own journal, and in Laszlo's keynote volume for the work of our group, *The New Evolutionary Paradigm,* I outlined the enormous benefits for moving chaos theory on beyond the hot house of mathematics into the vast open field of the social and life sciences.

In short I became, and was even heralded as the one who first saw and called for the new field for social science that has seen an enormous increase in use since then.

With Fred Abraham, Allan Combs, Stanley Krippner, and Sally Goerner, I became a co-founder of the multinational Society of Chaos Theory and Life Sciences.

This became SCTPLS for our acronym—which should defy any attempt to find an easy way to say it.

"Scoot please," if forced to it?

With respect to the exciting causes we can only say, as when speaking of so-called spontaneous variations, that they relate much more closely to the constitution of the varying organism, than to the nature of the conditions to which it has been subjected.

Darwin, *The Descent of Man*

SIX
FROM BIOLOGICAL TO CULTURAL EVOLUTION

Here was this book, this rambling wonder of *The Descent of Man*, which for over 100 years had been routinely probed by the fierce guardians of the language, methodologies, and mindset of traditional biology—as well as by the fierce guardians of the humanities. Here for the first time I saw the width and depth of the split in mindset that sent our species off in the wrong direction.

To go straight to the point, here was the endemic failure of science to discern and work with Darwin's move on from biological evolution into the wide new open territory of *cultural* evolution.

At this point in my investigation, as if right on schedule, came the stage in the traditional murder mystery where victims and burials begin to cumulate. One has suspicions but must wait for confirmations. As Freud and Jung first described the process, one must wait for the mysterious force within the unconscious mind to consolidate clues into some over-riding insight that links one thing to another over all.

It's where I came to see that up to now the work of most of science in this direction had been whistling in the wind—as if you had employed a detective to solve a murder mystery, then found he'd only been frittering the time away swapping tales with the good old boys in the corner saloon.

Affecting everything else I report in this book is the over-riding story within our story that began with the last year and the death of George Romanes—the young man so close to Darwin during his final decade that he became known as Darwin's disciple.

It was to Romanes that Darwin left all his notes on their ten year venture beyond biology into the psychology, and nascent systems science, of what Darwin called "the higher agencies." Romanes the wealthy disciple who went on to become a major British psychologist and important evolution theorist in his time.

It was Romanes who, in lamenting the prevailing obsession on natural selection to the exclusion of everything else, first called Darwin's famous successors "Neo-Darwinians."

Going blind and paralyzed, here is what Romanes wrote as he lay dying of cancer desperately trying to finish his book *Darwin and After Darwin.* In haunting detail this book foreshadowed much of what over the following century came to be. Was it only coincidence, or further evidence of what seemed to be trying to kill everything that might threaten the deadly Status Quo? For the book that Romanes desperately struggled to complete briefly blipped into publication in 1892, then went out of print for nearly 100 years.

Why "not only do the Neo-Darwinians strain the teachings of Darwin; they positively reverse these teachings," Romanes wrote.

..."so greatly have some of the Neo-Darwinians misunderstood the teachings of Darwin that they represent as 'Darwinian heresy' any suggestions in the way of factors 'supplementary to' or 'cooperative with' natural selection."

To replace the living Darwin he had known, indeed openly worshiped, Romanes charged the Neos were manufacturing a new scientific pseudo-religion with pseudo-priests to exploit the dead. Being enshrined was a "scientific creed ... not a whit less dogmatic and intolerant than was the more theological one which it has supplanted."

In Romanes, I found, was the missing link between the Darwin of biological evolution and the Darwin who to a far greater extent than had been realized had departed his peers, the Darwin who reveled in the joint delight of his worldwide pen pals exploring new territory.

It wasn't just what we've seen became the hot, new scientific territory of chaos, complexity, and self organizing theories. Back of and beyond this Darwin had gone on to probe the implications of brain and mind involved in his move beyond *biological evolution* to map the territory of *cultural evolution*.

It's vital to keep in mind that biological evolution for our species essentially ended with our arrival between 100,000 and 200,000 years ago. It became up to us to do something wonderful with the powerful new capacity for *choice of the future* that opened to our species, which Darwin brought to life in both science and story.

In science a pivotal move in this direction was publication in 1974 of Howard Gruber's book *Darwin on Man*. Howard had been a professor of mine during my years gaining the crucial PhD. At the time we students had laughed at the word going around. Professor Gruber was said to be at work on something so remote from psychology, and everything else that mattered on tests, as the quaint old Darwin.

But in Gruber's highly praised and national award-winning book I first saw the anchoring perception that in Darwin we were and are looking at two halves for his theory. There is the well known first half—but also a long buried second and completing half.

It was and is right there page after page in the buried rest of Darwin. For example in this—one of many such quotes—note how in speaking of "sympathy," "kindness," and "defending one another," Darwin was ranging beyond what became hard core NeoDarwinism.

> With mankind, selfishness, experience, and imitation, probably add, as Mr. Bain has shown, to the power of sympathy; for we are led by the hope of receiving good in return to perform acts of sympathetic kindness to others; and sympathy is much strengthened by habit.
>
> In however complex a manner this feeling may have originated, as it is one of high importance to all those animals which aid and defend one another, it will have been increased through natural selection; for those communities, which included the greatest number of the most sympathetic members, would flourish best, and rear the greatest number of offspring.

Here Darwin is clearly pointing toward how concern for others came into the picture. In other words, he is pointing toward the core need for morality, of which we've seen Darwin wrote 92 times, versus only twice for "survival of the fittest."

But also in this passage there is this. Who was this "Mr.Bain," to whom Darwin so often turns for support in writing of the impact of sympathy, imitation, and habit not on biological but rather cultural evolution.

Digging behind the name I found this was the first of 16 times Darwin refers to the work of Alexander Bain—the Scotch moral philosopher, social activist, author of the *Manual of Mental and Moral Science*, and a pioneering developer of the fields of social psychology, comparative psychology, and developmental psychology.

This was a revealing source, for Bain was considered the great psychologist of his time. Another of his stature was Pierre Paul Broca, considered the great pioneer in the science of the brain in his time, to whom Darwin refers 18 times.

Psychology, brain science—in Bain and Broca we glimpse the size of Darwin's move in two directions thereafter mainly ignored by the biologists pioneering evolution theory. Like the biblical "cloud no bigger than a man's hand on the horizon," we begin to more clearly see the size of his move from his celebrated theory of biological evolution to his long hidden prototheory of cultural evolution.

We begin to see the more sweeping, all-embracing shift from the celebrated natural science of *Origin of Species* into what became the buried social science of *The Descent of Man*.

Like everything else in Darwin that eluded his dedicated and respectful successors, here again is the fundamental insight that slips by so swiftly that only by knowing what you are looking for can you find it.

> With respect to the exciting causes we can only say, as when speaking of so-called spontaneous variations, that they relate much more closely to the constitution of the varying organism, than to the nature of the conditions to which it has been subjected.

Here is another answer to the question of what happened to the rest of Darwin. Natural selection is still in the picture—and still foundationally powerful. But now the rest of what shapes the wonder of cultural as well as biological evolution comes into play.

With way too much packed within a single line, it was sure to be missed both then and now. But by taking time to unpack it, I found we can jump considerably ahead in understanding.

Darwin is telling us that within "the constitution of the varying organism"—that is, *what was already there,* what *biological* evolution had already established within our species—is the power within ourselves to shape ourselves.

He is saying the wonderful human capacity for self-organizing can be far more important in shaping us than environment—that is, "the nature of the conditions to which it has been subjected."

It is as though the first half of Darwin's theory tells us that in climate change and comparable factors the chopper of natural selection is the force from outside ourselves that up to a point we must give in and adapt to. But here is the twist that opened to our species the wonder of the choice of the better or the agony of the worse future. Defying adaptation is the path through *cultural* evolution that wags a flag toward the better world.

In other words, in the imbalanced relation of male to female, the shackles of race, the clamp of custom, the demand of history, on and on, natural selection has become cultural selection at our species' level.

And behind it all is the force of love and the moral sense that Darwin worked so hard to get across. There is the shove that rather than give in drives us to overcome and prevail over whatever we feel is hateful rather than loving and wrong rather than right.

Yes this is complex, yes this is hard to follow. But this is the complexity that until late in the game made it easy for science to ignore *and thereby suppress the pivotal importance of Darwin's shift from biological to cultural evolution.*

In this and the next chapter is the hidden history of the split in our understanding of evolution, which led to and continues to lock in the hidden battle between the Gatekeepers and the Gatebreakers on which it looks like the fate of our species depends.

SEVEN
GATEKEEPERS AND GATEBREAKERS

Unless given a reason to become involved in the ups and downs of evolution theory, for scientist and lay man and lay woman alike, we tend to think of evolution as something of great importance that presumably someone else is looking after for us.

But the deeper I dug into the mystery of what buried the rest of Darwin the more I found the mess that actually existed *and still exists* in both the theory and the reality of evolution. Born out of the burial of the higher order rest of Darwin, here again was the clash of a New versus an Old paradigm, one band of us the Gatekeepers, the other the Gatebreakers.

We've come to know the Gatekeepers through the social, economic, and in particular political impact, pro and con, of three battling subgroups. Here again the sense of new victims for the standard murder mystery comes to mind. In this case we'll look at the unhappy interaction of the traditional Neo-Darwinians, the Super Neos, and the Creationists.

The traditional Neo-Darwinians are the generally amiable custodians of standard Neo-Darwinian evolution theory. Out of the early feeding frenzy over what was or wasn't what Darwin had in mind—of which we've seen his dying disciple Romanes despair and decry— this is the Grand Synthesis the great Julian Huxley and other leading biologists built to settle the matter in 1942.

It's important to know what they did—and didn't do. They

kept Darwin's primary theory of *natural selection* interacting with *variation*. They added Gregor Mendel's long ignored discovery of the powerful cross-generational factor of *genes*. But they gutted Darwin's concept of *variation*, which for Darwin had been natural selection's pivotal partner, and then shucked the rest.

The rest they dumped is what we'll spend the rest of this book uncovering. But regarding variation, what could be the problem with attaching that seemingly trivial little word *random* to Darwin's unadorned concept of variation? What could be wrong with transforming plain unadorned *variation* into *random variation*?

Wasn't the fact that much of evolution involved things that happened with seemingly no rime or reason—just collisions of unknowns that happened to bump into each other and thereafter for a while stuck together to either fight or love one another?

The problem was that by making all of evolution nothing more than one big bag of accidents the Neos ruled out the big thing for Darwin that made all the difference between sense and senselessness. For Darwin variation opened the way for the self-organizing process of human intervention that for better or worse drives us backward or ahead.

The Neo-Darwinians half-shut our species out of the process by replacing Darwin with the half-right idea of *random* variation —which in a rare outburst of exasperation and muted rage Darwin had roundly damned as the idea that evolution was exclusively ruled by "blind chance."

> The birth both of the species and the individual are equally parts of that grand sequence of events that our minds refused to accept as the result of blind chance. The understanding revolts at such a conclusion.

Lockstep at all levels, this brief compilation became the great Neo-Darwinian Synthesis, which for the favored theorists, writers, and delighted publishers of enormously profitable textbooks and "best sellers," still governs the official theory of evolution.

The Super Neos are the late-20th century pioneers of sociobiology and evolutionary psychology who set out to re-create all fields of social science in their own image. Pouncing on the failure of the NeoDarwinians to move on up the ladder from biological to cultural evolution, their goal was in the right direction, but the result was a radical diminishing of species that for a time became the most popular player in the game of evolution.

I won't attempt to substantiate that cryptic statement in short hand. I must refer you to the story of what happened that I have tried to untangle in chapters eighteen, nineteen, and twenty.

The Creationists tried to overturn and trash the game by denying evolution in any form other than literally by *deus ex machina.* That is, God in effect becomes the celestial gambler who owns the saloon and set in place and now in every detail runs the game. (See chapters ten and seventeen).

I knew how the Gatekeepers felt and still feel because I have been there. Like most scientists over all the years between the death of Darwin and recovery of the rest of his theory, for a long time I identified exclusively with the Gatekeepers.

It was what you automatically did as a scientist if faced with the need to relate your own work in any way to evolution. You dropped a thrice-blessed Gatekeeper book or two that you likely hadn't read into the references, then rallied to defend the great Neo-Darwinian synthesis against the vile Creationists. But with our species' over 100,000 year investment in evolution now facing nuclear wipe out, and formation of our group to try to head it off, I had joined the upstart world of the Gatebreakers who we'll meet in person and in groups throughout the rest of this book.

The Gatekeepers have a long, vital, and rightfully celebrated heritage in science and story. But in probing for what buried the rest of Darwin's theory, I found the buried story, neglected science, and frustration periodically alternating between rage or despair of the ***Gatebreaker*** heritage.

The Gatebreaker Heritage

It would take many books to make up for lost time and do it justice, but much is told in just two parables.

The Saga of Gatebreaker Suppression began, as we've seen, with the anguish, haunting death, and 100 year out of print oblivion for Darwin's disciple, George Romanes. My hope is that the glimpse I provide will inspire someone to dig into and write a book moving out from the agony of this central figure into a riveting portrait of the future that could have been ours. The other is the parable of Julian Huxley—grandson of Thomas Huxley, the two-fisted champion known as Darwin's bulldog, and founder of the fabled X Club of scientific support when Darwin was still considered only the engaging but lowly travel writer of *Voyage of the Beagle.*

As we've also seen, it was Julian Huxley who first wrote of, and in 1942 with twenty prestigious collaborators, developed, the deservedly great Neo-Darwinian Synthesis that still governs the Gatekeepers and thereby a crucial chunk of our species' mind.

But then, so swiftly, wholly, and conveniently forgotten, came this. For I found the story that went quickly in and out of both scientific and mass mind.

It is of how seventeen years later, in 1959, armed with the enormous prestige of being both the great scientist and action-oriented first Director of ENESCO, Julian Huxley tried to fill the hole in the mind of our species with a vital expansion beyond the sacred Synthesis into what by now was fairly screaming for attention.

Shifting from being co-founding leader of the Gatekeepers to a shocking new role as champion for the Gatebreakers, seizing every platform he could think of with a relentless outpouring of papers, talks and books, Huxley hammered at his peers with his powerfully reasoned concept of *psychosocial evolution.*

This he saw as the vital fully human, progressive action- and morally-oriented bridge from biological into cultural evolution, and from natural to social science. Driven by a vision of how we could expand and update his Grand Synthesis, he felt we could nudge our species to grow up on and out of its wayward adolescence into responsible maturity.

At first he was massively resisted. Then ignored. Then finally he was ostracized by many of his peers. For in his enthusiastic introduction to the powerful *Phenomenon of Man,* by the great visionary anthropologist and priest Teilhard de Chardin, Julian

Huxley dared suggest that science could benefit from a new look at its lock step rejection of the old hat factor of spirituality.

De Chardin, he wrote, "has forced theologians to view their ideas in the new perspective of evolution, and scientists to see the spiritual implications of their knowledge."

And so, as they say, the fat was in the fire.

EIGHT
SURF RIDING THE WAVES OF CHANGE

Cold shouldered by the Gatekeepers of evolution theory, evoking the buried vision of George Romanes and Julian Huxley, the Gatebreaker heritage came surf riding into history atop four waves of change that still pound at the Gate demanding entry.

One of them Abraham Maslow called the "third wave"—the impact of Freud being first, the impact of Behaviorism second, the third wave being the global spread of humanistic psychology and the vast human potentials movement.

Next was the spread of *evolutionary systems science.* Founded by Ludwig von Bertalanffy, Quaker futurist Kenneth Boulding and significant others, *systems science* broke through the feudal boundaries of all the fields of science, which had been chopped into separate baronies, to show them how to work together toward over-riding higher ends.

Most powerful, however, were two forces that, as with Darwin's scattered attempt to complete his theory, sought to bring the halves of an unbalanced world into a healing alignment. One was the vast cross-ocean rolling spread of thousands of years of Eastern spirituality into Western philosophy and psychology. The other was the fervent rise, global gatherings, and the "not going to take it any more" drive of the world wide women's movement.

Love, sex, community, moral sense, spirituality—suppressed in Darwin, re-emergent now in the higher order rest of his theory—all take on vast new meaning in the work of the great surf riders of the waves of change in our time.

I think of Dacher Keltner with his remarkable Greater Goodness Center at UC-Berkeley. George Lakoff with *Moral Politics* to wake the better angels of our political selves. Mihaly Csikszentmihaly, *The Evolving Self,* co-founding the new field of a morally Positive Psychology. Frans de Waal and the reassurance of our moral roots in the *Good Natured* ape.

As we saw earlier, David Sloan Wilson with the vital bridge between science and spirituality in *Darwin's Cathedral* and *Does Altruism Exist* to end the sacred reign of the Selfish Gene. Stanley Krippner and the indomitable push on into the beckoning reality that lies beyond the blind bias against the frontier of the paranormal.

And then in so many ways, most important of all: the insistent voice of all the women who over centuries have called for and are now driving the revolution in which the higher order rest of Darwin's theory was buried—that is, once again, the liberation of the bond of love, the clasp of community, the guidance of the moral sense and the comfort of spirituality.

Hazel Henderson's global fight for a moral economics to shame business to become truly better. Darcia Narvaes with her zoom in on child raising as the key to evolution and stellar leadership for the new field of moral psychology. Nel Noddings who fought for moral education while raising ten children. Barbara Marx Hubbard's experiment in bettering ourselves through conscious evolution ... and many, many more working to shore up our species in its endangered drive to fulfill itself.

What impels these special people? What are the personal experiences, motivations, reasons, functions, and goals that drive Gatebreakers such as these to rise each day keyed up to push on—for this is what I came to know of them.

They dared to dream the big dreams and didn't shrink from battle. In fact, matching research I knew of that confirmed this, they tend to relish going up against whatever calls for a fight to change. Of the surfers who set out to ride the waves of change I write of three I came to know best.

For riders of the wave of evolutionary system science: General Evolution theorist and founder of our group, Ervin Laszlo.

For riders of the wave blending Eastern spirituality with Western science: charismatic philosopher Ken Wilber, founder of Integral Theory.

Then in Part II, for a prototype riding the wave of the women's revolution: cultural evolution theorist Riane Eisler and her multilevel theory of cultural transformation.

Ervin Laszlo

Ervin began to amaze his elders as a child prodigy— he played his first piano concerto with the Budapest symphony orchestra at age nine. He went on to become a world-circling concert pianist, then gradually between concerts, stints of learning and teaching at the Sorbonne and Yale, he became a systems philosopher, pioneering general evolution theorist, and leader within the emerging field of multidisciplinary systems science.

What drove him? I found he was driven in two directions vital for advancing evolution—one out into the far reach of science, the other to prod science into action on the escalating pile of our world's most daunting problems. He became Research Director of a United Nations program and a major study of goals for world development for the prestigious Club of Rome. With the launching of our group and his own think tank, the Club of Budapest, he was probing for ways to get our species back on track in evolution.

The titles of a few of more than fifty books tell the action story. *Goals for Mankind: A Report to the Club of Rome. You Can Change the World. Strategy for the Future. The Choice: Oblivion or Evolution.*

Titles similarly track his development of scientific theory. *Evolution: The Grand Synthesis* laid out his general systems theory. *The Systems View of the World* expanded systems science into a larger world view. With *The Continuity Hypothesis,* he moved on to an understanding of the mind-blowing aspects of advanced physics unrivaled among evolution theorists.

His theory of quantum-vacuum interaction (QVI) emerges out of what seems empty space. Physicists calculate a single cubic centimeter of so-called empty space actually contains only a tiny portion of a gigantic body of still incomprehensible unfathomed energy.

Laszlo probes how we may tap into this exotic possibility shaping many aspects of our lives. In *Science and the Akashic Field: An Integral Theory of Everything* he links the Quantum Vacuum field to the Hindu Akasha—the ancient belief that everything that has ever happened, or is happening now, is in effect stored in a vast memory bank out of which our past, present, and future was and is being formed.

It is a theory that invites Gatekeeper skepticism but could explain much that still baffles science — particularly if one may add to it Darwin's long buried exploration of the evolutionary primacy of love.

"I acknowledge my role and responsibility in evolving a consciousness of all-embracing love in me, and by example in others around me," Laszlo writes in his late book *The All Embracing Love Declaration*.

"I have been part of the aberration of human consciousness in the modern age, and now wish to become part of the evolution that overcomes the aberration and heals the wounds inflicted by it.

"This is my right as well as my duty, as a conscious member of a conscious species on a precious and now critically endangered planet."

As time went by for our work to shuck "survival of the fittest" and update the crazy quilt of theory to something pointed toward a better world, I became increasingly fascinated by another work. It first shocked and then was fervently ignored by the Gatekeepers. As if with the steady beat of drums its devotees were proclaiming a new socio-spiritual-scientific path to the better world.

Within our group Allan Combs was the first to fall under its spell. A psychologist and neuroscientist by training, Alan had, as I've indicated, joined Fred Abraham and me in merry times advancing chaos theory. He was a comfortable bearded chunk of a man who, when startled, looked like a great white owl jogged out of some deep woodland thoughts.

To bridge the gap between the Gatekeepers and what to them was forbidden territory, Allan wrote a brilliant portrait of the rising importance of the study of *consciousness.* First published in 1995, *The Radiance of Being* became a classic for this then spanking new field.

In *Radiance* Allan showed how the focus on consciousness had emerged out of the "integral yoga" of Indian visionary Sri Aurobindo, the "integral structure" of French philosopher Jean Gebser, and now the far-ranging then new"integral theory" of the rambunctious American philosopher Ken Wilber.

Among much else what *Radiance* brought to life was how in contrast to what tended to be the other world spirit of Aurobindo and the sober scholarship of Gebser, Wilber was advancing a muscular, playful, provocative, and, above all, wide ranging *action-oriented* theory of evolution.

It looked, you might say, like putting wheels and a horn to the then new field of consciousness for grappling with how our brains intersect with the mystery of mind in shaping our evolution and future.

Ken Wilber

In the same year that *Radiance* was published, 1995, Wilber's flagship volume, the 831 page *Sex, Ecology, and Spirituality* came out, and for a while Wilber and Combs made a joyful splash in the sodden world of the status quo.

Earlier I'd read Wilber's book *The Atman Project.* In what seemed to me and many others a work of genius, first published in 1980, Wilber had pulled together a staggering range of works integrating Eastern spirituality with both Western science and Western spirituality.

Here was not just another static academic map of possibilities. Symbolized by the moral-action-oriented Hindu vision of the Atman who acts to advance evolution, all this was caught up within what thousands of years of suffering and aspiration have tried to tell us of the drive within each of us of the ideal, of a higher ultimate form—in other words, the inescapably absolute requirement for *action* to better both ourselves and the world around us.

This became the unifying *leit motif* and mission for a steady stream that became 45 books that gained a fervent global readership in many languages. The books further roused the critics who loaded Wilber with the sure fire label for generating big book sales: that golden word controversial!

In a classic case of what founding sociologist Max Weber called the routinization of the charisma, Wilber built a global network of professional practitioners, counselors, and co-developers to apply his Integral Theory to self development, business, politics, warfare, both personal and economic depression, spirituality, and much else troubling our species.

Again a few book titles tell the story. *Integral Psychology. Integral Meditation. Integral Spirituality. The The Integral Vision. The Integral Mission. A Brief History of Everything. A Theory of Everything. Quantum Questions. Grace and Grit. The Marriage of Sense and Soul*— and what in 2017 became the monumental *The Religion of Tomorrow,* and *Trump and a Post-Truth World.*

As I began to uncover the lost rest of Darwin, I found myself increasingly prodded toward Wilber's work.

Darwin's great advance I began to see, as if carved in stone, had come from how in a more simple time he had been able to transcend enough of the complexities to grasp the vital sense of evolution as an integrated whole.

Now out of an immense increase in the complexities, it seemed clear that both Laszlo and Wilber had set out to try to do the same for our time. Out of this venture had come Laszlo's formation of our group and work toward his Akashic Field Theory. Now here was the exotic rise of Wilber's work and the tempting mystery of Wilber's *Quadrants.*

I found how key factors in my reconstruction of Darwin's completed theory can be arranged in terms of drivers and stages for a hypothetical wedding of biological and cultural evolution. Add them up for Darwin and you get nine stages and seven drivers for a total of 16 in all.

Now for Wilber's Quadrants (see brief guide here in

Resources and Reflections) what struck me was that in comparison with Darwin's 16, for Wilber each of the lines that form the four quadrants contains 13 elements, for a count of 52 in all.

Given this hint of mathematical assurance that I might be on the right track—that is, expansion of Darwin's 16 to Wilber's 52—I began to look for a mirroring of Darwin's lost insights in both Wilber's work and integral theory more generally.

Could I find what we've seen so far in Darwin's startling prefiguring of self-organizing processes and chaos theory?

Digging for the Darwin-Wilber Connection

For Wilber's work the most comprehensive single source is *Sex, Ecology, and Spirituality*—or the invaluable brash shorter *A Brief History of Everything.* Long before the Gatekeepers even bothered to consider admitting the strange "new" idea of self-organizing processes to the official pantheon for evolution theory Wilber had turned to Laszlo for this quote.

> The new sciences dealing with these 'self-winding' or 'self-organizing' systems are known collectively as the sciences of complexity.

Further exhibiting his early comprehension of their significance, I found that Wilber wrote of the spread of these "new sciences" in terms of pioneering authorities— chief among them, Laszlo, Prigogine, Varela, and Abraham, the stars for our advanced research group.

Then came this fascinating finding. The index for Wilber's *Sex, Ecology, and Spirituality* further tells us that Wilber had already ventured on and widened his search to include "self-actualization, self-adaptation, self-dissolution, self-esteem, self-preservation, self-renewal, self-knowledge, self-transcendence, and self-transformation."

This was certainly impressive, but could early Wilber also contain chaos theory?

Long before the Gatekeepers would even mention such a thing existed, in 1995, again quoting Laszlo, Wilber wrote, "In recent years chaotic behavior has been discovered in a wide variety of natural systems. An entire discipline has sprung up within dynamic systems theory devoted to the study of the properties of chaotic attractors ... properly known as chaos theory."

By now I was excited by what looked like the case of two bold thinkers with theories that might interlink and support one another.

So far the tracks for self organizing processes and chaos theory did indicate something very much worth exploring. But would the trend hold for the rest of Darwin, which we'll explore in the rest of this book in Part II?

I was eager to go on, but then it happened. Like biting down on a rock in a box of candy I was jolted to a stop. It was in an area vital to my own work in which Wilber was just plain wrong.

Eventually, with personal contact and accelerating wonderment, it faded, but at the time it was like the puncturing of a hot air balloon. Further pursuit of the Darwin-Wilber connection became one of those things we tell ourselves we'll get around to later—and never do.

But it was hard to resist his scope as an offbeat thinker, his playfully barbed social critiques, and his hypnotic flair as a writer.

Though leery, I kept lightly in touch.

> When the thunder roars, do you not hear your Self? When the lightening cracks, do you not see your Self? When clouds float quietly across the sky, is this not your very own limitless Being, waving back at you?

> Evolution does not isolate us from the rest of the Kosmos, it unites us with the rest of the Kosmos: the same currents that produced birds from dust and poetry from rocks produce egos from ids and sages from egos.

This was nice, but more captivating was the structure of an action-oriented theory in tune with the rest of Darwin that seemed to be emerging in Wilber.

> We move from part to whole and back again, and in that dance of comprehension, in that amazing circle of understanding, we come alive to meaning, to value, and to vision: the very circle of understanding guides our way, weaving together the pieces, healing the fractures, mending the torn and tortured fragments, lighting the way ahead—this extraordinary movement from part to whole and back again, with healing the hallmark of each and every step, and grace the tender reward.

> At the Integral stages of development, the entire universe starts to make sense, to hang together, to actually appear as a uni-verse—a "one world"—a single, unified, integrated world that unites not only different philosophies and ideas about the world, but different practices for growth and development as well.

And there was this sober insight into the global fact of potential nuclear annihilation and the jolt of the grim reality of Islamic terrorists erupting like poisonous toadstools on the green lawn of the better world.

> If some sort of reconciliation between science and religion is not forthcoming, the future of humanity is, at best, precarious.
>
> There is arguably no more important and pressing topic than the relation of science and religion in the modern world. Science is clearly one of the most profound methods that humans have yet devised for discovering truth, while religion remains the single greatest force for generating meaning. Truth and meaning, science and religion; but we still cannot figure out how to get the two of them together in a fashion that both find acceptable.
>
> These two enormous forces—truth and meaning—are at war in today's world ... And something sooner or later has to give.

PART II

RECOVERING THE REST OF DARWIN'S THEORY

Or she may accept, as appearances would sometimes lead us to believe, not the male which is the most attractive to her, but the one which is the least distasteful.
Darwin, *The Descent of Man*

TEN
SEX

Of the "lost"factors that I found within the well buried, even paved over, rest of Darwin, in what lies ahead two aspects share meanings in common. One, it could be said, was their status as fresh victims within the course of our hypothetical murder mystery. The other was the fact all were Gatekeeper rejects that Gatebreakers fervently embraced and pursued.

We'll begin with a victim that on at least one level Keepers and Breakers were heartily in tune. Behind the clues, neither lost nor undetected but rather battered into bewilderment, was Darwin's theory of sexual selection.

Within *Descent*'s 791 pages I found 106 references to sexual selection in human evolution and 348 references to sexual selection in other species.

This would seem more than enough to settle the matter. But as Darwin's successors set out to refine and improve the majestic sprawl of what he left them, what originally seemed clear in Darwin became a cause for the pitched battles among his successors that became known as the Darwin Wars.

Here was a new answer to the question of what buried the rest of Darwin. Sexual selection was a pet of his, a cherished insight of major importance to Darwin. But in digging into what happened I found the twisted dynamic of major concern, which over this and the next eight chapters unfolds. For while on one hand the Gatekeepers were obviously advancing the evolution of our species, on the other hand one could see how their radical reduction of Darwin to only the first half of his theory was diminishing our species and driving our beeline course toward destruction.

I found, as can be seen in quotes that lie ahead, what seems to have been the Gatekeeper effort to bury the anti-macho, indeed even proto-feminist implications, of what Darwin actually wrote about sexual selection.

There was more to it. For the other tactic for Darwin's successors was an attempt to collapse sexual selection into natural selection as the handy answer for everything—as before Darwin, God had been the all-purpose answer.

They turned to the gene in the search for the ultimate answer. For three decades the lowly monk Gregor Mendel's epochal experiments with yellow and green pea plants had been ignored. Then in 1900, within just two months, independently from one another, three eminent scientists rediscovered Mendel and the magic of the gene. This became an immense advance for evolution theory. But just as with natural selection originally, once again the focus on the gene became obsessive.

It was like the famous quote by humanistic psychology pioneer Abraham Maslow: "If you only have a hammer, you tend to see every problem as a nail."

The result of what became a feeding frenzy for thousands of studies by biologists is captured in this salacious example from Evolution 101, online for UC-Berkeley.

> It's clear why sexual selection is so powerful when you consider what happens to the genes of an individual who lives to a ripe old age but never got to mate: no offspring means no genes in the next generation, which means that all those genes for living to a ripe old age don't get passed on to anyone! That individual's fitness is zero.
>
> Selection makes many organisms go to extreme lengths for sex: peacocks maintain elaborate tails, elephant seals fight over territories, fruit flies perform dances, and some species deliver persuasive gifts. After all, what female Mormon cricket could resist the gift of a juicy sperm-packet? Going to even more extreme lengths, the male redback spider literally flings itself into the jaws of death in order to mate successfully.

This is vivid, engaging, all true. But here is another case of how burial of the rest of Darwin devastated both science and society. We see how the overwhelming focus on the gene, as a matter of biological evolution, blanked out recognition of the force of *gender*, as a matter of *cultural* as well as biological evolution.

Step by step, out of hints and clues along the way, I came to see the crucial full significance of this observation.

Here we can see how Darwin began to probe sexual selection by defining two basic forms.

Darwin on Sexual Selection

The sexual struggle is of two kinds; in the one it is between individuals of the same sex, generally the

> males, in order to drive away or kill their rivals, the females remaining passive; whilst in the other, the struggle is likewise between the individuals of the same sex, in order to excite or charm those of the opposite sex, generally the females, which no longer remain passive, but select the more agreeable partners.
>
> Sexual selection has been treated at great length in this work; for, as I have attempted to shew, it has played an important part in the history of the organic world.

This is the broad general picture that he begins to qualify and modify. The male and masculinity, as we have known it, is overwhelmingly dominant—but as if "on little cat feet" a subtle difference enters the scene.

> *The Power of Love vs. the Power of Battle*
>
> In the lower divisions of the animal kingdom, sexual selection seems to have done nothing: such animals are often affixed for life to the same spot, or have the sexes combined in the individual, or what is still more important, their perceptive and intellectual faculties are not sufficiently advanced to allow of the feelings of love and jealousy, or of the exertion of choice.
>
> When, however, we come to the Arthropoda and Vertebrata, even to the lowest classes in these two great

> The males are almost always the wooers; and they alone are armed with special weapons for fighting with their rivals. They are generally stronger and larger than the females, and are endowed with the requisite qualities of courage and pugnacity...
>
> This surprising uniformity in the laws regulating the differences between the sexes in so many and such widely separated classes, is intelligible if we admit the action of one common cause, namely sexual selection.

Now are we only seeing what is happening at the prehuman level? Or are we also beginning to see how it's mirrored in all the funny, fearful or endearing instances that come to mind at the level of development for our species?

Are we seeing all the boys we knew who spruced up with tie, flowers, and guitar to go courting? Or all the girls with twinkly dresses, sparkling ear rings, and deft foot to catch your eye at dances?

What automatically comes to your mind as you read on?

> *The Power of Charm, Love Song, and Dance*
>
> The development, however, of certain structures—of the horns, for instance, in certain stags—has been carried to a wonderful extreme; and in some cases to an extreme which, as far as the general

> conditions of life are concerned, must be slightly injurious to the male. From this fact we learn that the advantages which favoured males derive from conquering other males in battle or courtship, and thus leaving a numerous progeny, are in the long run greater than those derived from rather more perfect adaptation to their conditions of life.
>
> We shall further see, and it could never have been anticipated, that the power to charm the female has sometimes been more important than the power to conquer other males in battle.

It's as if we're beginning to see what is there before us in print within the blend of two different levels of evolutionary emergence. Yes, Darwin is writing about sexual selection at the prehuman level. But at the same time—and this likely both in his own mind originally, and certainly in our minds today—we are leaping ahead to an inner slide show of memories at our later level of emergence.

What becomes a battle to the death at the prehuman level becomes a matter of outraged or despairing name calling at our level for girls, or the tender knock at the door with flowers and a box of candy for boys.

Up to this point the idea that we move out of selection at the prehuman biological level into fulfillment at the human cultural level had only been another denatured abstraction. So far it was like a deflated tire that needed air to be useful. But now I was seized by the wonder of the trans-evolutionary unfolding of the living, yearning, wanting reality of life at our level.

Aren't we looking ahead to see ourselves in the astonishing

variety of personalities, communities, roles, and plot lines that our memories open to us at our level of cultural evolution?

And through it all we're seeing how Darwin began to develop the case he made for the female of the species that quietly but persistently began to upset the macho apple cart.

Are we looking at Darwin's exploration of a higher order missing link? In this case the hidden crossover point from biological to cultural evolution?

The Power and Function of the Sham Battle

In the case of Tetrao umbellus, a good observer goes so far as to believe that the battles of the male "are all a sham, performed to show themselves to the greatest advantage before the admiring females who assemble around; for I have never been able to find a maimed hero, and seldom more than a broken feather."

I shall have to recur to this subject, but I may here add that with the Tetrao cupido of the United States, about a score of males assemble at a particular spot, and, strutting about, make the whole air resound with their extraordinary noises. At the first answer from a female the males begin to fight furiously, and the weaker give way; but then, according to the first answer from a female the males begin to fight furiously, and the weaker give way; but then, according to Audubon, both the victors and the vanquished must either then exert a choice, or the battle must be renewed.

So, again, with one of the field-starlings of the United States (Sturnella ludoviciana) the males engage in fierce conflicts, "but at the sight of a female they all fly after her as if mad."

Are we looking at Darwin's exploration of a higher order missing link? In this case the hidden crossover point from biological to cultural evolution?

The Plot Thickens

That the males of all mammals eagerly pursue the females is notorious to every one. So it is with birds; but many cock birds do not so much pursue the hen, as display their plumage, perform strange antics, and pour forth their song in her presence. The male in the few fish observed seems much more eager than the female; and the same is true of alligators, and apparently of Batrachians.

Throughout the enormous class of insects, as Kirby remarks, "the law is, that the male shall seek the female."

One parasitic Hymenopterous insect forms an exception to the rule, as the male has rudimentary wings, and never quits the cell in which it is born,

> whilst the female has well-developed wings. Audubon believes that the females of this species are impregnated by the males which are born in the same cells with them; but it is much more probable that the females visit other cells, so that close inter-breeding is thus avoided.
>
> We shall hereafter meet in various classes, with a few exceptional cases, in which the female, instead of the male, is the seeker and wooer.

Haven't we been seeing how step by step, out of of all the tasty separate items at the prehuman level, Darwin was uncovering the vast smorgasbord out of which evolution chose freedom of choice for the behavioral feast at our level?

Isn't this the special gift to our species that we have yet to learn to appreciate? And that now we must do so or join the dinosaurs?

> Woman seems to differ from man in mental disposition, chiefly in her greater tenderness and less selfishness ... Man is the rival of other men; he delights in competition, and this leads to ambition which passes too easily into selfishness.
>
> Darwin, *The Descent of Man*

TEN
MORE ON SEX

Should I be questioned about my use of "mangling" to characterize what Darwin's successors did to sexual selection, I must apologize but cannot help straying to why I feel the word at this point seems apt.

A fascinating giant mechanism for my, as well as other ancient childhoods, was the moaning monster they called the mangle. Kept in the back room and animated on wash days, the mangle was used to iron sheets and shirts. At one fell swoop it could roll them flat as the proverbial pancake—as was the fate for the Neo-Darwinian mangling of the gene and gender levels of sexual selection.

This thought triggered a memory that automatically I felt had no place here. For the dominant canon for Gatekeeper science has been to at all costs avoid anything that might smack of the personal—or humor.

The word "I" is also to be scrupulously avoided. Instead you must automatically hide behind the safe and proper word "we."

But mindful of all the years science has been an exclusive good old boys club, with rampant mangling of sexual selection, I decided to chance it.

For what came back to me was that deep voice out of nowhere that originally invited me to come to the secret meeting in Budapest, which led to the formation of our research group, which had tried to beat back the specter that ever since 1965 has haunted our time.

Part of me was greatly excited by the call. To fly behind the Iron Curtain during the Cold War to discuss a mysterious high level project that might help head off nuclear annihilation—who wouldn't jump at the chance. But I had by then decided that the women's movement was the only thing that might save us. So while one part of me was eager to say yes, another part of me sternly warned this could be just one more puffed up, chest pounding, all male exercise in rearranging, as was said, the deck chairs on the Titanic.

I decided to try to get the person and work of a particular woman included.

"None of us are bringing their wives," I was told.

This was back in the days when women were so automatically excluded from anything that even began to smack of access to power that only rarely could the lone exception be included.

"But she's not my wife," I said. "We're not married. She just happens to be doing the most advanced work I know of on the big issues you say we'll be addressing. It's essential to understand how we got into the global fix we're in, and how to get out of it, and I think she's got a pretty compelling answer."

This I felt was no exaggeration, as I was the first to read what came out of her ten year struggle to research and write her first big book.

The first time they met without me. I felt awful—had I lost my one slim chance to make a difference in what happened to our poor battered species and planet? But they came back a second time.

"Bring her," they said.

They gingerly edged around her at first, and for a second or two fell silent whenever she dared to speak, but Riane Eisler soon became welcomed as an exceptionally pretty one of the boys.

And within three years her book, *The Chalice and the Blade,* was hailed by the crusading anthropologist Ashley Montagu as "the most important book since Darwin's Origin of Species."

Riane Eisler

Hers was a haunting, raw, and in terms of both history and human evolution deeply meaningful story. In the same year that Sigmund Freud fled Vienna from the Nazis, 1938—during the infamous Kristalnacht when Jewish stores and synagogues were burned and men hauled off for extermination in concentration camps—little seven-year-old Riane Tennenhaus had seen her father kicked downstairs by the Gestapo to haul him off to certain death.

She had also seen her mother stand up to the armed men who broke into their home and win him back. They hid him in the attic until they could escape. Then with the odds heavily against them they fled to Havana, Cuba—one of the only two places left in the world still admitting desperate Jews fleeing Nazi Europe.

She was there in Havana on the lonely thrust into the sea of the Malecon watching when the Cubans shut the port against more Jews and turned back the St.Louis. She saw the "Ship of the Damned," with nearly 1000 desperate refugees, forced to sail on with by now frantic hope for rescue by by America—only to be turned back by the U.S. government, forcing their return to Europe and for many death in the concentration camps.

After the Allied victory and the discovery and liberation of the gaunt, shriveled, hollow-eyed remnant still left alive, there in Havana she saw the first newsreels revealing the horror of the Holocaust that had wiped out 6,000,000 Jews including her beloved grand parents, uncles, aunts, and cousins throughout Europe.

This became the cauldron of memory that drove her to find out how and why this happened—and most urgent and crucial, what could be done to keep it from ever happening again.

She got to America, married, had children, gained the grounding in systems science and the law degree that gave her a way to make a difference in this world. She became a notably innovating leader in the late 20th century rise of the global women's movement. She gave up much of everything else for the years of research out of which came her new theory of human cultural evolution, and books in 36 languages with hard won solutions for the worst of threats to human survival.

The Chalice and the Blade: Our History, Our Future powerfully documents and grounds her work in the pattern for 35,000 years of our species' cultural evolution—the long early rise of a peaceful, gender-equalitarian and highly creative partnership system system ... ended by a violent shift to 5,000 years of an authoritarian, male-ruled domination system ... out of which we are involved in the new rise of partnership system values, goals, and peace in every aspect of our lives.

Sacred Pleasure: Sex, Myth, and the Politics of the Body showed how the conflict. of partnership ways with dominator ways shapes our sex, love, political and spiritual life.

Tomorrow's Children showed how schools tuned to the partnership way of life could help prevent the nightmare of violence over and over agaln.

The Power of Partnership is a practical guide for shifting from dominator to partnership ways of life.

The Real Wealth of Nations shows how her new model of caring economics can help speed the shift from the dominator model for economics now driving business, government and our species toward destruction, to the humane, caring, and sustainable world still within our grasp.

In the descriptions I've been using to give a brief touch of life to everybody, how should I describe her?

For many acquainted with her work, or the dark history of our species out of which it emerged, there is an inspiring, one of a kind sense of history and fulfilment about her work and story.

She was her own person, set apart from all others, who for me brought to mind the ethereal beauty of the actress Hedy Lamar. Both were Viennese, Jewish, gifted with exceptionally high I.Q.s, both linked in different ways to the Holocaust.

As for Darwin, along with many of the rest of us today born male, he's become known as the prototypical male chauvinist. But aren't we seeing—and ahead will see more of—the little anti-macho digs he snuck into the long buried rest of his theory?

Could it be that, next here, this traitorous quote is more evidence of how, why, and who buried the uppity rest of Darwin?

Could it be one doesn't have to be a systems scientist to suspect the 5,000 year old paradigm of male domination had a hand in it?

Woman as a Moral Being

Woman seems to differ from man in mental disposition, chiefly in her greater tenderness and less selfishness; and this holds good even with savages, as shewn by a well-known passage in Mungo Park's Travels, and by statements made by many other travellers.

Woman, owing to her maternal instincts, displays these qualities towards her infants in an eminent degree; degree; therefore it is likely that she would often extend them towards her fellow-creatures.

As Whewell has well asked, "who that reads the

touching instances of maternal affection, related so often of the women of all nations, and of the females of all animals, can doubt that the principle of action is the same in the two cases?"

Man is the rival of other men; he delights in competition, and this leads to ambition which passes too easily into selfishness.

> ... an increase in the number of well-endowed men and an advancement in the standard of morality will certainly give an immense advantage to one tribe over another.
>
> Darwin, *The Descent of Man*

ELEVEN
COMMUNITY

Behind the clues that pointed to the mangling of sexual selection, I also found the ferocious charge and counter-charge of the battle over community selection, or "group selection," as it is known today.

Of this Darwin has much less to say. But in what he says, two things hang on in mind. One is his conclusion about what in the end most matters in life. The other is again the immense loss our species suffered from burial of the rest of Darwin.

> *Darwin on Community Selection*
>
> Ultimately our moral sense or conscience becomes a highly complex sentiment originating in the social instincts, largely guided by the approbation of our fellow-men, ruled by reason, self-interest, and in later

times by deep religious feelings, and confirmed by instruction and habit.

It must not be forgotten that although a high standard of morality gives but a slight or no advantage to each individual man and his children over the other standard of morality gives but a slight or no advantage to each individual man and his children over the other men of the same tribe, yet that an increase in the number of well-endowed men and an advancement in the standard of morality will certainly give an immense advantage to one tribe over another.

A tribe including members who, from possessing in a high degree the spirit of patriotism, fidelity, obedience, courage, and sympathy, were always ready to aid one another, and to sacrifice themselves for the common good, would be victorious over most other tribes; and this would be natural selection.

At all times throughout the world tribes have supplanted other tribes; and as morality is one important element in their success, the standard of morality and the number of well-endowed men will thus everywhere tend to rise and increase.

We have now seen that actions are regarded by savages, and were probably so regarded by primeval man, as good or bad, solely as they obviously affect the welfare of the tribe—not that of the species, nor that of an individual member of the tribe.

> This conclusion agrees well with the belief that the so-called moral sense is aboriginally derived from the social instincts, for both relate at first exclusively to the community.

Seems to make sense, doesn't it?

The nature and power of community is what first Darwin, and then the revolutionary Prince Peter Kropotkin, one of the great natural scientists of his time, explored as the matrix of *mutual aid* in his great classic simply and powerfully titled *Mutual Aid*.

It is the cherished subject of much of the whole field of anthropology, for example in the works of Hortense Powdermaker, Ruth Benedict, Margaret Mead, and Ashley Montagu. It is what Emile Durkheim, Max Weber, Robert Bellah and the long list of the Chcago-born classic community studies, in which I became heavily involved, pursued in sociology. In psychology it is what Milton Rokeach, with whom I worked, Abraham Maslow, Muzifer Sherif, the great authors of The Authoritarian Personality—which I and many others still consider the single most important research project of the 20th century, again horribly relevant in our time—all worked.

Within our group I found my earlier love of community studies come to life in endless fascinating ways. For example, how often work that advances us emerges out of a group of people who seem to have been thrown together rather haphazardly, as might seem to be the case for our group, but who out of the bonding of like interests become a notably creative community.

You feel appreciated, nurtured, protected—it's the feeling of a home in an alien world.

Who could argue with the idea that through the agency of community, now called *group selection,* we can sometimes accelerate the drive toward a better world by joining our hands and cause with the hands and cause of others?

Indeed, why on earth would anyone in their right mind set out to destroy it as a major factor bearing on evolution? Yet this appealing observation for Darwin was picked up for ostensibly further development by the Super Neos for whom it became the holy cause for pitting *individual selection* against *group selection.*

Speedily seizing for themselves the sacred role of the Gatekeeper, "Super Neos," I decided, was the best fit name for the end of the 20th century salvo of the sociobiologists and evolutionary psychologists who, under the claim of being the newest and truest of Darwin's successors—as I write of in chapters eighteen and nineteen—took over much of the American, and thereby global, publishing industry.

The main difference between individual selection and group selection seemed to be easy to understand. One involved the selective power inherent in people generally and the other the honed and more specific joint power of the bond of community. Yet within the burgeoning cohort of the Super Neos the idea of group selection was like tossing red meat to a pack of wolves.

Here's what the brilliant Super Neo Steven Pinker had to say of what became the snarling clash of claim and counter claim.

> Group selection has become a scientific dust bunny, a hairy blob in which anything having to do with "groups" clings to anything having to do with "selection." The problem with scientific dust bunnies is not just that they sow confusion; ... The problem is that it also obfuscates evolutionary theory by blurring genes, individuals, and

> groups as equivalent levels in a hierarchy of selectional units; ... this is not how natural selection, analyzed as a tgismechanistic process, really works.
>
> Most importantly, it has placed blinkers on psychological understanding by seducing many people into simply equating morality and culture with group selection, oblivious to alternatives that are theoretically deeper and empirically more realistic.

The counter-blast for group selection came with publication of *Unto Others: The Evolution and Psychology of Unselfish Behavior* by philosopher Elliot Sober and biologist David Sloan Wilson.

This became the fight that fired Wilson to break out of the tunnel visioned impasse for biologists to tackle the big questions for open-minded evolution theorists. He became the leader of the group selection and multi-level selection movement, and co-founder and president of an increasingly influential online Evolution Institute, out of which he launched a stream of influential books (see bibliography) expanding the territory of Darwin's higher order focus on the moral sense (chapters fourteen and fifteen in this book) and the no longer so shocking role of spirituality in evolution (see chapters sixteen and seventeen).

Regarding the Super Neo bashing of group selection, I feel further ompelled to ask this.

In the light of what we face in the struggle of democracy versus the frightful shift to solve the problem of governing with autocracy or tyranny, isn't *group* selection why we vote?

The case for democracy? To get together and act together for progressive impact on our lives, our history, and our evolution?

It would seem that only by working overtime could you force individual selection and group selection apart to create opponents to fight one another.

Like ourselves in most any normal endeavor—say housework, or child raising, or on the job—don't both work by themselves as well as together?

Isn't this what Darwin was saying?

The Community of Delight

It's evanescent, hard to capure, but I find the idea of the bonding power of a community of delight in creative group ventures, whether successful or not, trggered by memories of the playful inner world of science that bubbled to the surface in our group.

Behind Ralph Abraham's massive learning and pioneering works, for example, was his colorful early life as an academic hippie. He'd apparently gone through the sixties requisite mix of experiments with communes, transformative drugs, and spirituality—I particularly relished the picture of his early experiment in living as in effect a temporary disciple in a cave with a guru in India.

His older brother Fred, who comes in and out of our story, had been and even still was even more the wild man. A fellow psychologist whom I came to know and work with in the development of a vital phase of chaos theory, Fred pioneered translation of basic concepts and theories for psychology into the language of this baffling new field of theory.

I cherished Fred's tale of how he would paddle in a canoe out to the middle of the lake in Vermont, where he lived, to serenade the wilderness with a convoluted version of the trumpet that looked like a squashed tuba.

He further reveled in the sideline that widened our experience of community out into the delight of others in our group. For Fred and his weird trumpet it was weekly sessions that of last count were still continuing with Fred as a hot performer in a jazz band composed of what I pictured as wild old University of Vermont academics.

Within our group, prior to his shift to science, CIIS professor of integral studies Monty Montuori, basked in memories of his jazz band in London and current marriage to Kitty Margolis, a top ranked jazz singer.

In far off New Zealand Ray Bradley, whom we'll meet in our chapter on love, was behind the beat, clang, and rattle of the drum combo in still other jazz bands.

I myself played both jazz and classic in my skit as "the only man in the world who can play Beethoven's Ninth on a one inch long harmonica."

I tell of colorful details like this not only because I want to make the buried science of Darwin's "lost voyage to a better world" come to life in this book. I tell of it because this is a glimpse behind the old facade in science, which one seldom gets, into the worldwideworldwide community of delight that out of free-wheeling and fun-loving lives like these chaos and complexity theory were emerging.

Within the titanic struggle over who and what is going to shape the future, it seems to me this generally unknown and under-valued delight is a vital aspect of the creativity that drives progressive social action we need to understand and put to use as a bedrock matter of species survival

> At times the male will chase the female all over the aviary, then go to the bower, pick up a gay feather or a large leaf, utter a curious kind of note, set all his feathers erect, run round the bower and become so excited that his eyes appear ready to start from his head ... until at last the female goes gently towards him.
>
> Darwin, *The Descent of Man*

TWELVE
LOVE

That Darwin in *The Descent of Man* could write 95 times about love and this be ignored for over 100 years is obviously shocking. But this was only the tremor for an earthquake. For despite the fact the word love was there trying to call attention to itself 95 times there was only a single entry for love in the index. And after over 100 years of being used but unchanged, in by now countless millions of copies, once again, *as of this writing, this loveless index is still in use in all editions of Descent worldwide.*

Why should this matter?

Typically a book's index is of little interest to the general reader. It's generally skipped without a qualm. But for the scholar the index is the sacred *sine qua non* for all who write the books, teach the students, and influence the leaders in every field bearing on whether we go forward, are checked in place, or driven backward in evolution.

A general reader will automatically read a book through from first to last page—but not the scholar, or student, or anyone else forced to cram a lot in overnight.

With few exceptions we glance at the dust jacket to see who recommends it, quick scan the table of contents, then zip eagerly to the index to see whether or not we, or our friends, or our own tidy little world of hot personal interests and references is in it.

In other words, if it's not in the index it's as good as dead and gone. Add up how this went on for over a century and you can see another reason why the rest of Darwin was condemned to rattle around in what increasingly seems to be the hole in the head of modern mind.

Foreshadowing what over time became science, over 2000 years ago the colorful Greek philosopher we earlier mentioned, Empedocles, made a case for love and strife (or love versus war) as the two prime drivers of evolution.

In 1893 the man whom many consider the greatest American philosopher, Charles Sanders Peirce, developed a compelling case for the drive of *evolutionary love.*

In a fascinating leap of insight that integrates the known with the unknown rest of Darwin, Peirce reasoned that evolution was driven by three forces. *Mechanical Necessity* and *Fortuitous Variation* were his terms for Darwin's interaction of natural selection and variation. But to these two, Peirce added *Creative Love* as the third driver. And then most powerfully evocative was this. He said that *all three* were embraced and at work within the over-riding thrust of *Evolutionary Love* as the prime driver of evolution.

Likewise is the wedding of science and spirituality in the work of Pierre Teilhard de Chardin. An anthropologist on one hand famous for his discovery of Peking Man and on the other his parish life as a Catholic priest, at some time prior to his death in

1955 de Chardin wrote of love in a way remarkably mirroring both Darwin and Peirce.

Suppressed by the church during de Chardin's lifetime is this passage in which de Chardin mirrors Darwin's similarly suppressed higher order completion of his theory.

> Considered in its full biological reality, love—that is to say, the affinity of being with being—is not peculiar to man. It is a general property of all life and as such it embraces, in its varieties and degrees, all the forms successively adopted by organized matter.
>
> In the mammals, so close to ourselves it is easily recognized in its different modalities: sexual passion, parental instinct, social solidarity, and so forth ...
>
> This is no metaphor; and it is much more than poetry, de Chardin concludes.

And so the mystery deepens. For how could all this in Darwin be dropped out of what was called his theory, as if it had never existed?

Within our group three of us were working to develop a new understanding of the role of love in evolution. For Riane Eisler love, of which she wrote extensively in her book *Sacred Pleasure,* had a key place in her cultural transformation theory.

As Darwin gingerly poked toward it and Eisler massively confirmed it, on one hand was the prevailing power of the brutal "combat and ravish" *domination* systems model prevailing in male-to-female and female-to-male relationships, family life, child raising, and most other areas of our lives.

On the other hand was the countering influence of the long subordinated more gentle, loving, egalitarian and peaceful *partnership* systems model now in particular re-emerging through the thrust of sanity over the past 300 years.

For myself there was a similar place for love in my moral transformation theory—which in addition to three books of love poems eventually became ten mainly still unpublished books including this one. And then there was and is the brilliant sociologist Ray Bradley.

A wiry New Zealander, with a romantic flare of white hair like Karl Pribram, Ray, as noted earlier, was also an irregular drummer with a jazz band for relaxation. He was also a talented painter. In science he got underway in the 1960s with an impressive update for sociology's co-founder Max Weber's classic study of the enchantment of *charisma*. Ray had shown how through the power of love, interacting with the power of control, charisma operated in the social and sexual arrangements of forty six communes throughout America.

Working within the heady mix of chaos, complexity, and self-organizing theory, often in tandem with Karl Pribram, Ray was in effect updating Empedocles with what several of us felt could be headed toward a major contribution to evolution theory.

Most intriguing to me was Ray's connection over 100 years to the lost Darwin.

Darwin on Love

It is not difficult to imagine the steps by which the notes of a bird, primarily used as a mere call or for some other purpose, might have been improved into a melodious love song.

> We will confine our attention to the higher social animals and pass over insects, although some of these are social and aid one another in many important ways.
>
> The social animals which stand at the bottom of the scale are guided almost exclusively, and those which stand higher in the scale are largely guided, by special instincts in the aid which they give to the members of the same community ...
>
> Such animals are always ready to utter the danger-signal, to defend the community, and to give aid to their fellows in accordance with their habits; they feel at all times, without the stimulus of any special passion or desire, some degree of love and sympathy for them; they are unhappy if long separated from them, and always happy to be again in their company.

Insects, social animals at the bottom of the scale, issuing danger signals, feeling unhappy on parting—it's hard for most of our self-absorbed higher species to think of anything at the "lower level" as love. But Darwin's was the revolutionary focus on where love began in sex and in widening and deepening became the sublime force binding us self to self and others.

The Season of Love

With the great majority of animals ... the taste for the beautiful is confined, as far as we can judge, to the attractions of the opposite sex. The sweet strains poured forth by many male birds during the season of love, are certainly admired by the females, of which fact evidence will hereafter be given.

If female birds had been incapable of appreciating the beautiful colours, the ornaments, and voices of their male partners, all the labour and anxiety exhibited by the latter in displaying their charms before the females would have been thrown away; and this it is impossible to admit.

In other species the males become brighter than the females and otherwise more highly ornamented, only during the season of love. The males sedulously court the females, and in one case, as we have seen, take pains in displaying their beauty before them.

Can it be believed that they would thus act to no .purpose during their courtship? And this would be the case, unless the females exert some choice and select those males which please or excite them most.

Bartram describes the male alligator as striving to win the female by splashing and roaring in the midst of a lagoon, "swollen to an extent ready to burst, with its head and tail lifted up, he springs or twirls round on the surface of the water, like an Indian chief rehearsing his feats of war." During the season of love, a musky odour is emitted by the submaxiliary glands of the crocodile, and pervades their haunts.

Even with the most pugnacious species it is probable that the pairing does not depend exclusively on the mere strength and courage of the male; for such males are generally decorated with various ornaments, which often become more brilliant during the breeding season, and which are sedulously displayed before the females.

The males also endeavour to charm or excite their mates by love-notes, songs, and antics; and the courtship is, in many instances, a prolonged affair.

Mr. Bartlett has described to me the behaviour of two chimpanzees, rather older animals than those generally imported into this country, when they were first brought together. They sat opposite, touching each other with their much protruded lips; and the one put his hand on the shoulder of the other. They then mutually folded each other in their arms.

Afterwards they stood up, each with one arm on the shoulder of the other, lifted up their heads, opened their mouths, and yelled with delight.

Darwin, The Expression of Emotions

THIRTEEN
MORE ON LOVE

No doubt it will still be said these are just quaint little stories about animals. What is the relevance to us? Or to science?

The point is Darwin writes of love not in the customary string of abstractions for scientific or scholarly discourse, which tend to flatten life into the corpse of the butterfly pinned beneath glass. He reports of what he and his delightful band of world-circling pen pals found of love in the concrete living presence—that is, in life in its reality before abstraction.

It is this reality of love in its unfolding up through time from possibly the earliest species to ourselves that Darwin writes of. Long seen in this way in the expression of spirituality, I feel we are looking at another instance of Darwin's pioneering for a science slow to catch up.

Every one must have noticed how miserable horses, dogs, and sheep are when separated from their companions, and what strong mutual affection horses and dogs, at least, show on their reunion.

It is curious to speculate on the feelings of a dog, who will rest peacefully for hours in a room with his master or any of the family, without the least notice being taken of him. But if left for a short time by himself, he barks or howls dismally.

The love of a dog for his master is notorious; as an old writer quaintly says, "A dog is the only thing on this earth that luvs you more than he luvs himself."

I have myself seen a dog, who never passed a cat who lay sick in a basket, and was a great friend of his, without giving her a few licks with his tongue, the surest sign of kind feeling in a dog.

Pelicans fish in concert. The Hamadryas baboons turn over stones to find insects, and when they come to a large one, as many as can stand around, turn it over together and share the booty.

Social animals also perform many little services for each other. Horses nibble, and cows lick each other on any spot that itches. Monkeys search each other for external parasites. And Brehm states that after a troop of the *Cercopithecus griscoviridis* has rushed through a thorny brake, each monkey stretches itself on a branch, and and another monkey sitting by "conscientiously" examines its fur, extracting every thorn or burr.

Many animals ...certainly sympathize with each other's distress or danger. This is the case even with birds. Captain Stansbury found on a salt lake in Utah an old and completely blind pelican. This pelican was very fat, so it must have been well fed for a long time by its companions.

Captain Stansbury also gives an interesting account of the manner in which a very young pelican, carried away by a strong stream, was guided and encouraged in its attempts to reach the shore by half a dozen old birds.

Mr. Hewitt states that a wild duck, reared in captivity "after breeding a couple of seasons with her own mallard, at once shook him off on my placing a male Pintail on the water. It was evidently a case of love at first sight, for she swam about the new-comer caressingly, though he appeared evidently alarmed and averse to her overtures of affection.

From that hour she forgot her old partner. Winter passed by, and the next spring the Pintail seemed to have become a convert to her blandishments, for they nested and produced seven or eight young ones."

In further pursuit of what buried Darwin's delightful paean to love, I found another answer to the question who did it. For after thousands of years of being celebrated by great poets, religious visionaries, philosophers, artists, composers and ourselves in every day life, how could love be so quickly strangled and dumped out of sight in science?

In part the villain was the anti-anthropomorphism mindset that seized biology after Darwin. That is, in the drive to ground the theory of evolution in hard, cold "facts" and the blessing of mathematics, anthropomorphism—that is, attributing human attributes to animals, plants, or anything else prior to us on the ladder of evolution—became one of the worst of sins.

Love? Awful. Further sinful was any use of anecdotes. One must pin everything down with the scientific exactitude of experimental or statistical control. Thus wherever Darwin used the squishy word love—or the humorous little stories he delighted in—became *verboten*.

On a hunch I took at look at *Origin of Species.* And sure enough I found *Origin* loaded with anthropomorphism and anecdotes from beginning to end.

So why could the first half for Darwin's theory be snapped up, hailed, and widely put to use by both science and society, but the nascent second and higher order completion of his theory be dumped?

One further question and two probable answers.

In contrast to all that Darwin wrote about "love among the animals," why was there so little about "love among the humans" in *The Descent of Man*?

One must keep in mind that Darwin's was the Victorian period when even the legs of the furniture had to have skirts.

Replacing God with natural selection had already been more than enough shock to sensibilities—add sex among humans and face a trial to rival the fate of poor Oscar Wilde. Neither Darwin nor the world was ready for the shock of Freud, Havelock Ellis, or—perish the thought!—Wilhelm Reich on *The Function of the Orgasm*.

A deeper reason, however, further reveals why Darwin was in effect buried alive. It was his place in history as the last of the great naturalists. From a peak in the climb, as it were, he could look out and down at nature as a wedded whole before the split became a gulf separating social science from natural science and biological from cultural evolution.

Darwin did not think of himself as a psychologist, as his brilliant cousin Francis Galton became. Or as a psychologist, sociologist, systems scientist, and ethicist as well as a philosopher, as his friend the redoubtable Herbert Spencer was. Scientifically, what went on at the much more complex level for the development of our species was alien territory to Darwin. Yet with a depth, breadth, economy and accuracy that eluded both Galton and Spencer he boldly pushed himself on to an amazing achievement.

Only now can we begin to see how over 100 years ago Darwin reached ahead and did so well with our species out of his uncanny level of understanding earlier species—as shown by this case of a little bird world domestic drama that points ahead to the delightful repeat performance among ourselves that perpetuate our species.

The Bower Birds

But the most curious case is afforded by three allied genera of Australian birds, the famous Bower-birds—no doubt the co-descendants of some ancient species which first acquired the strange instinct of constructing bowers for performing their love-antics.

The bowers, which, as we shall hereafter see, are decorated with feathers, shells, bones, and leaves, are built on the ground for the sole purpose of courtship, for their nests are formed in trees.

Both sexes assist in the erection of the bowers, but the male is the principal workman.

So strong is this instinct that it is practised under confinement, and Mr. Strange has described the habits of some Satin Bower-birds which he kept in an aviary in New South Wales.

"At times the male will chase the female all over the aviary, then go to the bower, pick up a gay feather or a large leaf, utter a curious kind of note, set all his feathers erect, run round the bower and become so excited that his eyes appear ready to start from his head; he continues opening first one wing then the other, uttering a low, whistling note, and, like the domestic cock, seems to be picking up something from the ground, until at last the female goes gently towards him."

If we turn then from love among his beloved animals to the ways and wiles of love among ourselves, Darwin adds this in *The Expression of Emotion in Man and Animals.*

> No emotion is stronger than maternal love; but a mother may feel the deepest love for her helpless infant, and yet not show it by any outward sign; or only by slight caressing movements, with a gentle caressing movements, with a gentle smile and tender eyes. But let any one intentionally injure her infant, and see what a change! How she starts up with threatening aspect, how her eyes sparkle and her face reddens, how her bosom heaves, nostrils dilate, and heart beats; for anger, and not maternal love, has habitually led to action.

And finally, there is this from *The Descent of Man:*

> Although man, as he now exists, has few special instincts, having lost any which his early progenitors may have possessed, this is no reason why he should not have retained from an extremely remote period some degree of instinctive love and sympathy for his fellows.
>
> A man who possessed no trace of such instincts would be an unnatural monster.

Society could not go on except for the moral sense, any more than a hive of Bees without their instincts.

Darwin, *The Descent of Man*

FOURTEEN
MORAL SENSE

If Darwin writing 95 times about love with but a single entry in the index of *The Descent of Man* is shocking, what of the fact he wrote 92 times of moral sensitivity *and no one seemed to know it?*

It seems impossible, incredible, what can one say—the virtual annihilation of Darwin as a moral theorist seemed to have stopped just short of being a sweeping *fait accompli*.

In hundreds of books over by now close to a century, I found only four notably aware of the moral Darwin—chief among them the remarkable book by the University of Chicago's noted historian of science Robert J. Richards, which we'll look at in the next chapter. As otherwise this giant omission is unbelievable, to avoid having to labor the point, here's just one instance of what opened my eyes to what deserves to be called the hole in the head for modern mind.

In the years prior to my search behind the clues for what had buried the sparkling rest of Darwin, I spent many months of research to identify pioneers in the study of the *moral sense* for my book *The River and the Star: The Story of the Great Scientific Explorers of the Better World*.

Out of lives and works familiar to authorities in moral philosophy and social science, I wrote extensively of Immanuel

Kant, Herbert Spencer, Marx and Engels, Emile Durkheim, Jean Piaget, Lawrence Kohlberg, Carol Gilligan, as well as others I'd personally known and worked with.

So effectively had the moral Darwin been buried that in not even one of their books did I find a single mention of Darwin as in any way significantly concerned with moral evolution and theory.

Earlier, in chapter eleven, we briefly explored the feeling of a home in an alien world Darwin explored as the matrix of *mutual aid*.

Within our group the very nature of our mission —to build a theory of evolution that might provide a guide to the better rather than worse future—indicated the level of moral sensitivity prevailing among us through Laszlo's selection of our members.

More specifically, besides the investment in this direction of my partner, Riane Eisler, and myself, was the cosmic probe of the noted astrophysicist Eric Chaisson.

Eric was a genial, sandy-haired, bespectacled fellow with what became a long time Harvard faculty and observatory involvement. Along one track Eric became a pivotal builder of the famous Hubble Telescope, which from 353 miles out from earth explores ever deeper and further into space. Along another track, he wrote some good books to reflect what happened all the way from the explosion of cosmic evolution to the rise of moral evolution within "the life era." But particularly striking was how it was as if he and the Hubble telescope were joined by some cosmic mystery to Immanuel Kant's famous quote linking the moral law within us to the stars.

> "Two things fill the mind with ever new and increasing admiration and awe, the more often and steadily we reflect upon them: the starry heavens above me and the moral law within me."

Over the years no other quote had so deeply and often moved me. Looking back now as I move into my nineties, I find it haunting to see how much the moral drive of Laszlo's vision, and the spread and comfort of that rare band of minds, led me to uncover the lost Darwin.

It was like the watering of a seed long ago implanted and waiting within me. For out of me there poured this shaking of a fist at everything that conspired to diminish our species.

And here, once again, is the lonely quote from where it sat ignored for over 100 years. For there it sits, on page [531], within the first of three paragraphs that end what Darwin clearly labeled his General Summary and Conclusion for *The Descent of Man.*

> Important as the struggle for existence has been and even still is, yet as far as the highest part of man's nature is concerned there are other agencies more important.
>
> For the moral qualities are advanced, either directly or indirectly, much more through the effects of habit, the reasoning powers, instruction, religion, &c.,

Darwin's language may here and there seem quaint or the science dated. But here in all these quotes for the first time exhumed and gathered together in one place, is what surely must come to be recognized as of major importance.

Here, from page after page in *Descent*, are passages that reveal Darwin's long denied but rightful stature in going on beyond philosophy and religion to ground a new *science* of moral as well as general evolution in an everlasting great rock for eternity.

Darwin on the Moral Sense

I fully subscribe to the judgment of those writers who maintain that of all the differences between man and the lower animals, the moral sense or conscience is by far the most important.

This sense, as Mackintosh remarks, "has a rightful supremacy over every other principle of human action;" it is summed up in that short but imperious word ought, so full of high significance. It is the most noble of all the attributes of man, leading him without a moment's hesitation to risk his life for that of a fellow creature; or after due deliberation, impelled simply by the deep feeling of right or duty, to sacrifice it in some great cause.

Immanuel Kant exclaims, "Duty! Wondrous thought, that workest neither by fond insinuation, flattery, nor by any threat, but merely by holding up thy naked law in the soul, and so extorting for thyself always reverence, if not always obedience; before whom all appetites are dumb, however secretly they rebel ..."

This great question has been discussed by many writers. Mr. Bain gives a list of twenty-six British authors who have written on this subject, and whose names are familiar to every reader; to these, Mr. Bain's own name, and those of Mr. Lecky, Mr. Shadworth Hodgson, Sir J. Lubbock, and others, might be added, of consummate of consummate ability ... my sole excuse for touching on it, is the is the impossibility of here passing it over; and because, as far as I know, no one has approached it exclusively from the side of natural history.

The investigation possesses, also, some independent interest, as an attempt to see how far the study of the lower animals throws light on one of the highest physical faculties of man.

Of Selfishness

It was assumed formerly by philosophers of the derivative school of morals that the foundation of morality lay in a form of Selfishness; but more recently the "Greatest happiness principle" has been brought prominently forward.

Nevertheless, all the authors whose works I have consulted, with a few exceptions, write as if there must be a distinct motive for every action, and that this must be associated with some pleasure or displeasure. But man seems often to act impulsively, that is from instinct or long habit, without any consciousness of pleasure, in the same manner as does probably a bee or ant, when it blindly follows its instincts.

Under circumstances of extreme peril, as during a fire, when a man endeavours to save a fellow creature without a moment's hesitation, he can hardly feel pleasure; and still less has he time to reflect on the dissatisfaction which he might subsequently experience if he did not make the attempt.

Should he afterwards reflect over his own conduct, he would feel that there lies within him an impulsive he would feel that there lies within him an impulsive power power widely different from a search after pleasure or happiness; and this seems to be the deeply planted social instinct.

A distinct feeling that our impulses do not by any means always arise from any contemporaneous or anticipated pleasure, has, I cannot but think, been one chief cause of the acceptance of the intuitive theory of morality, and of the rejection of the utilitarian or "Greatest happiness" theory.

With respect to the latter theory the standard and the motive of conduct have no doubt often been confused, but they are really in some degree blended.

When a man risks his life to save that of a fellow-creature, it seems also more correct to say that he acts for the general good, rather than for the general happiness of mankind.

Thus the reproach is removed of laying the foundation of the noblest part of our nature in the base principle of selfishness ......

This was the man who was never considered a moral theorist, who was attacked by Creationists as the monster who brought on a disastrous rise in immorality, who was lampooned by cartoonists as a pretentious, ponderous, pseudo-scientific subhuman ape.

This was the man who for ten decades after his death, in hundreds of prestigious books on morals and morality, seemed to be connected with morality not even in a casual passing way.

FIFTEEN
MORE MORAL SENSE

The mystery further deepens. For after Darwin thought through, and wrote out, and even within his own lifetime saw it published in a book for a global readership, how could all this still be buried?

In the impassioned passages we've just read he refers to the mighty Mackintosh, John Stuart Mill, Immanuel Kant, and Marcus Aurelius—men considered giants in the study of moral evolution.

Sir James Mackintosh, whom Darwin considered the greatest influence on him during his formative teen years, was the then leading living authority on the anti-selfishness, moral sense aligned Scottish Enlightenment, of which I write in *Darwin in Love*.

Immanuel Kant, writing in the 18th century, is still considered the most influential of all moral philosophers. John Stuart Mill, writing of economics, politics, and the liberation of women, was considered the most influential of 19th century philosophers. Marcus Aurelius was the Roman Emperor who, wholly contrary to expectation, leaped nearly 2000 years ahead in stating the moral goals of our time and beyond. Yet Darwin who now looms as the pioneering giant of moral science remained unknown.

The mystery even further deepens if we look at the book that first opened my eyes to the revolutionary fact of Darwin as a moral theorist. This was Robert J. Richards' *Darwin and the Emergence of Evolutionary Theories of Mind and Behavior*.

Richards was and is the great historian of science and award-winning Darwinian scholar who, in pages of rare wit and keen insight, rediscovered Romanes and first unearthed the story of Darwin's long ignored status as a moral theorist, which I expand in this book. Repeatedly, with the power of an undeniable high level of scholarship, Richards revealed how the moral Darwin was blanked out by his successors.

In one of many examples of burials that follow the traditional pattern for a murder mystery, perhaps most striking is how two of the most famous fighters for Darwin's cause dumped the moral connection—as well as much of the rest of the "lost" Darwin!

In 1893, T.H.Huxley, the pivotal friend known as "Darwin's bulldog," founder of the famous X Club of crucial support for the fledgling Darwin, proclaimed you simply can't get morality out of the apparent fixation in cosmic home territory for what was already beginning to morph into the survival of the fittest/selfish genes mindset and theory.

"Of moral purpose I see not a trace in nature," he told the gathering of the great, at the Sheldonian Theatre in Oxford, for the supreme irony of what by then had become the prestigious Romanes Lecture. "That is an article of exclusively human manufacture," Huxley proclaimed, even as Romanes himself, riddled with cancer, speechless, lay dying.

Then in 2002, in the 1400 pages of his enormous last book, published posthumously, the best known Darwinian of his time delivered the *coup de grace.*

"As humans, we surely have a legitimate personal interest in our moral behavior," wrote Stephen Jay Gould, "but we cannot enshrine this property as occupying more than a tiny corner of nature."

Here was the immensely popular scholar, author of such delightful titles as *The Panda's Thumb, The Flamingo's Smile,* and *Bully for Brontosaurus.* And in drawing on his knowledge of

Darwin's works in *The Structure of Evolutionary Theory,* he further provides only three skimpy pages on why neither in Darwin nor evolution theory is there any significant place for morality.

This tells us much of the degree to which Darwin's successors either never read him. Or skimmed past the word moral without seeing it. Or ultimately, I believe, unconsciously blacked out whatever failed to fit within the firmly grounded first half of Darwin's discovery that was being twisted into the paradigm of Survival of the Fittest and Selfish Genes *uber alles*.

Darwin on Right versus Wrong in "Man" and Animals

It may well be first to premise that I do not wish to maintain that any strictly social animal, if its intellectual faculties were to becomes as active and as highly developed as in man, would acquire exactly the same moral sense as ours.

In the same manner as various animals have some sense of beauty, though they admire widely different objects, so they might have a sense of right and wrong, though led by it to follow widely different lines of conduct.

If, for instance, to take an extreme case, men were reared under precisely the same conditions as hive-bees, there can hardly be a doubt that our unmarried females would, like the worker-bees, think it a sacred duty to kill their brothers, and mothers would strive to kill their fertile daughters; and no one would think of interfering.

Mr. H. Sidgwick remarks, in an able discussion on this subject, "a superior bee, we may feel sure, would aspire to a milder solution of the popular question."

Judging, however, from the habits of many or most savages, man solves the problem by female infanticide, polyandry and promiscuous intercourse; therefore it may well be doubted whether it would be by a milder method.

It is to be hoped that the belief in the permanence of virtue on this earth is not held by many persons on so weak a tenure. Nevertheless, the bee, or any other social animal, would gain in our supposed case, as it appears to me, some feeling of right or wrong, or a conscience.

For each individual would have an inward sense of possessing certain stronger or more enduring instincts, and others less strong or enduring; so that there would often be a struggle as to which impulse should be followed; and satisfaction, dissatisfaction, or even misery would be felt, as past impressions were compared during their incessant passage through the mind.

In this case an inward monitor would tell the animal that it would have been better to have followed the one impulse rather than the other.

The one course ought to have been followed, and the other ought not; the one would have been right and the other wrong ...

> .. a belief in all-pervading spiritual agencies seems to be universal; and apparently follows from a considerable advance in man's reason, and from a still greater advance in his faculties of imagination, curiosity and wonder.
>
> Darwin, *The Descent of Man*

SIXTEEN
SPIRITUALITY

Surprise? Shock? Delight? Reactions will vary according to one's prior beliefs.

Thus we come to the anxiety of those hovering at the jumping off point from the spring board of science toward spirituality, or from spirituality toward science.

That is, the quandary of the scientist nudged by a sense of something of major impact at work beyond the prevailing proper limits of science, or the spiritually aligned nudged by a sense of some probably improper but essential need for grounding in science.

For over 100 years we've been given the picture of Darwin as the famous prime rejecter of a belief in God—which indeed he was. Further widely implanted is the idea that for Darwin religion and spirituality were God's partners in crime.

Another wave of books proclaiming the death of God and predicting the same for religion erupted out of the late 20th and early 21st century shift to the reign of the Super Neo, which we'll look at in chapters eighteen and nineteen.

But I found that if we look at the lost rest of Darwin with even halfway open eyes and mind, something considerably different from the prevailing assumption about what Darwin believed comes to light.

In keeping with the automatic indifference prevailing in scientific circles, within our group there was no open discussion of God, religion, or spirituality. However, privately among ourselves, as well as widely elsewhere, it was a different story.

Most fervent and increasingly influential was Riane Eisler's alignment to the rapidly growing women's spirituality movement, in which she became a cherished thinker following a pioneering involvement in the women's movement earlier. In both *The Chalice and the Blade* and *Sacred Pleasure* she also pursued the good and bad sides for religion and spirituality.

There was Ralph Abraham. Born of an early involvement with Hindu beliefs, his book *Chaos, Gaia, and Eros* was a zany mix of religion, spirituality, chaos theory, and environmental concern.

Francisco Varela, who died of hepatitis in 2001, was an ardent Buddhist and scientific adviser to the Dalai Lama.

There was Laszlo with his unique blend of new science and ancient spirituality in the theory of the Akashic Field we looked at in chapter eight.

As for myself, I was a Christian probably headed for the ministry until switched to journalism by World War II. Thereafter routed out of journalism into science, I now find myself, as an old man, point for point in tune with where the old Darwin arrived.

Darwin and God, Religion, and Spirituality

To understand what Darwin really and finally believed, one

must know something of the dilemma of Darwin's father on finding himself saddled with a teen age son woefully lacking in any sign of a promising direction in life.

To please his father Charles first tries medical school, but has to drop out because he is overwhelmingly repelled by the bloody brutality of what he is required to watch and do. Then came the big hope for both father and son as Charles moved on from Edinburgh to Cambridge.

Here he not only gained the degree that would qualify him for a guaranteed lifetime employment as a minister. He actually excelled in divinity studies, in final exams ranking 10th in a class of 178. Then came the voyage of the Beagle and the transformation of a prospective minister into the fierce vision of becoming a famous scientist.

On returning home, in his long unpublished early notebooks, appear the pivotal first insights into what became the biological foundation for the first half for his theory in *The Origin of Species.*

For example:

> . . . now my theory makes all organic being perfectly adapted to all situations where in accordance to certain laws they can live ... All flow from some grand & simple laws ... inconvenience! *extinction,* utter *extinction*!.. Let him study Malthus & Decandoelle. ... I look at every adaptation, as the surviving one of ten thousand trials ...

But also—and, significantly, at far greater length—are the notes that became the foundation in cultural evolution for the completion of his theory buried within *The Descent of Man*.

For example:

> May not moral sense arise from our ... strong instinctive sexual, parental, & social instincts, giving rise [to] "do unto others as yourself," "love thy neighbor as thyself."

This is the first of Darwin's three great leaps ahead in his Early Notebooks. This is the first statement of the historic insight, which, as we shall see, collapsed what became his wedding of biological and cultural evolution into joining the severed halves, which unfolded into the awesome, and indeed epochal completion of his theory of evolution..

Now for still another factor in the burial of the rest of Darwin—and this a big one: the impact of the Creationists.

Darwin and the Creation of Creationism

What we generally fail to realize about the Creationists is how great is the damage science brought on itself as the unwitting instigator, indeed virtual creator of Creationism.

Had the Creationists been confronted with the Darwin of love, the moral sense, and, as we're seeing in this and the next chapter, his long overlooked good word for the impact of spirituality on evolution, their power would have been deflated, even over time maybe ended.

Instead, the battle of Creationists versus Evolutionists has for over a century sopped up the energy in science that could have gone into advancing the theory of evolution in Darwin's original direction. Still more damaging is how over the same long span of time the battle of the Evolutionists versus the Creationists has sopped up most of the news coverage in America for anything to do with evolution. This has meant a huge loss of space in media that otherwise might have been used to uncork the relevant social and systems science the Gatekeepers had bottled up and set aside.

It's meant the fear of teaching anything but a tiny token bit on evolution to keep school boards and legislatures infiltrated by Creationists at bay.

According to a Gallop Poll in June of 2014, 40 percent of Americans believe in Creationism and 76 percent said they would wouldn't be upset if creationism were taught in their schools.

Only 22 percent—just two out of ten—said they'd be upset.

Historically, here midway through the 19th century, was a huge body of people suddenly confronted with a science in which one could find little of love and nothing of morality. So with gusto Creationism rose to fill the vacuum and try to drive us backward in evolution.

What would Darwin have said of this woeful alternative?

> The same high mental faculties which first led man to believe in unseen spiritual agencies, then in fetishism, polytheism, and ultimately in monotheism, would infallibly lead him, as long as his reasoning powers remained poorly developed, to various strange superstitions and customs.

SEVENTEEN
MORE SPIRITUALITY

Ahead lies the unexpected fascination of more of what Darwin wrote that for over 100 years most of his scientific successors automatically skipped over. But first we must consider what in terms of our familiar analogy might be called another accessory to murder determining how and why the rest of Darwin was buried.

Much has been made of the conflict between Charles and his beloved wife Emma on matters of religious belief. This tends to stick in mind as a portrayal of the big bad gulf between religion and science. But this seems to be another case of what pushed the rest of Darwin out of the picture for scientists and, even more devestating, for story tellers.

There was no place for the rest of Darwin in the no nonsense, "just the facts, mam" science being established by his successors. But if we look at the actual life behind the tale of Charles and Emma with some knowledge of the Unitarian faith in that time as well as ours, a different tale emerges.

Joined in their marriage, the Wedgewoods for Emma and the Darwins for Charles, were families both steeped in what then was the controversial departure from standard Christianity of the Unitarian faith. What I found was that what's been seen as the prototypical story of the pivotal clash between science and religion was actually an easily tolerated difference in schools of the Unitarian faith.

Within Darwin himself—as is clear in his Autobiography —four kinds of belief lived uncomfortably together, contending for supremacy at various points in his life.

There was the traditional Church of England or Anglican faith, toward which his studies at Cambridge were directed, the only sure funded source for a career as a minister in that time.

But in sharp doctrinal difference were two kinds of Unitarians—neither of which offered much hope for funding and a viable career. There were the moderates, who hung onto some of the traditional beliefs and rituals, and the radicals, who in essence rejected everything but a fervent commitment to moral action, or otherwise doing good in the world.

Closely intertwined with the radical Unitarians were the wholly non-church going free thinkers, of whom Darwin's beloved grandfather Erasmus, and his father, as well as Marx and Engels, were prime examples.

And in the middle of it all—caught between the clasp of his immense sensitivity to the feelings of others and the fierce sense of mission to solve the mystery of evolution—was Darwin, forced to try to juggle all this in mind.

Freed by Unitarianism to believe or not to believe in God, Emma was a believer who fervently retained the powerfully traditional Christian belief in an after life.

Charles, by contrast, believing neither in God nor the after life, chose to give up God and formal church-going religion. But he remained aligned to the Wedgewood and Darwin family code—that is, recognition of the function of a belief in spirituality as a practical step up in evolution while heartily rejecting the dogma, which for Unitarians was the curse hung round the neck of traditional belief.

The crunch point for Charles and Emma came with the death of their ten–year-old daughter Annie. The searing pain drove Emma to implore Charles for the comfort of believing they might all be rejoined in heaven after death. But in the wrench of honesty, this he could not do.

> *Darwin on God*
>
> There is no evidence that man was aboriginally endowed with the ennobling belief in the existence of an Omnipotent God.
>
> On the contrary there is ample evidence, derived not from hasty travellers, but from men who have long resided with savages, that numerous races have existed, and still exist, who have no idea of one or more gods, and who have no words in their languages to express such an idea.
>
> The question is of course wholly distinct from that higher one, whether there exists a Creator and Ruler of the universe; and this has been answered in the affirmative by some of the highest intellects that have ever existed.

Skeptics in his time, and ever since, have passed off this ringing endorsement for "God the Creator" as merely something Darwin threw in to disarm and hold off the rage of religious critics.

Undoubtedly, at least in part, this is true. But in the spirit of the real rather than fictional Darwin, this rings as more meaningfully his grateful tribute to the great philosophers and spiritual visionaries he studied at Cambridge, whom he cites often in his writings, as well as to all the ministers who remained close friends throughout his life

More of Darwin on God

The belief in God has often been advanced as not only the greatest, but the most complete of all the distinctions between man and the lower animals. It is however impossible, as we have seen, to maintain that this belief is innate or instinctive in man.

I am aware that the assumed instinctive belief in God has been used by many persons as an argument for His existence. But this is a rash argument, as we should thus be compelled to believe in the existence of many cruel and malignant spirits only a little more powerful than man; for the belief in them is far more general than in a beneficent Deity.

The idea of a universal and beneficent Creator does not seem to arise in the mind of man, until he has been elevated by long-continued culture.

Darwin on Religion

The feeling of religious devotion is a highly complex one, consisting of love, complete submission to an exalted and mysterious superior, a strong sense of fear, reverence, gratitude, hope for the future, and perhaps other elements.

No being could experience so complex an emotion until advanced in his intellectual and moral faculties to at least a moderately high level.

Nevertheless, we see some distant approach to this

state of mind in the deep love of a dog love of a dog for his master, associated with complete submission, some fear, and perhaps other feelings.

The behaviour of a dog when returning to his master after an absence, and, as I may add, of a monkey to his beloved keeper is widely different from that towards their fellows.

In the latter case the transports of joy appear to be somewhat less, and the sense of equality is shewn in every action.

Professor Braubach goes so far as to maintain that a dog looks on his master as on a god.

It is said that Bacon long ago, and the poet Burns, held the same notion.

Darwin on Spirituality

If, however, we include under the term "religion" the belief in unseen or spiritual agencies, the case is wholly different; for this belief seems to be universal with the less civilised races. Nor is it difficult to comprehend how it arose.

As soon as the important faculties of the imagination, wonder, and curiosity, together with some power of reasoning, had become partially developed, man would naturally crave to understand what was passing around him, and would have vaguely speculated on his own existence.

As Mr. M'Lennan has remarked, "Some explanation of the phenomena of life, a man must feign for himself, and to judge from the universality of it, the simplest hypothesis, and the first to occur to men, seems to have been that natural phenomena are ascribable to the presence in animals, plants, and things, and in the forces of nature, of such spirits prompting to action as men are conscious they themselves possess."

It is also probable, as Mr. Tylor has shewn, that dreams may have first given rise to the notion of spirits; for savages do not readily distinguish between subjective and objective impressions.

When a savage dreams, the figures which appear before him are believed to have come from a distance, and to stand over him; or "the soul of the dreamer goes out on its travels, and comes home with a remembrance of what it has seen."

In a like manner Mr. Herbert Spencer, in his ingenious essay in the 'Fortnightly Review, 'accounts for the earliest forms of religious belief throughout the world, by man being led through dreams, shadows, and other causes, to look at himself as a double essence, corporeal and spiritual. As the spiritual being is supposed to exist after death and to be powerful, it is propitiated by various gifts and ceremonies, and its aid invoked.

He then further shews that names or nicknames given from some animal or other object, to the early progenitors or founders of a tribe, are supposed after a long interval to represent the real progenitor of the tribe; and such animal or object is then naturally believed still to exist as a spirit, is held sacred, and worshiped as a god.

Nevertheless I cannot but suspect that there is a still earlier and ruder stage, when anything which manifests power or movement is thought to be endowed with some form of life, and with mental faculties analogous to our own.

The tendency in savages to imagine that natural objects and agencies are animated by spiritual or living essences, is perhaps illustrated by a little fact which I once noticed: my dog, a full-grown and very sensible animal, was lying on the lawn during a hot and still day; but at a little distance a slight breeze occasionally moved an open parasol, which would have been wholly disregarded by the dog, had any one stood near it.

As it was, every time that the parasol slightly moved, the dog growled fiercely and barked. He must, I think, have reasoned to himself in a rapid and unconscious manner, that movement without any apparent cause indicated the presence of some strange living agent, and that no stranger had a right to be on his territory.

The belief in spiritual agencies would easily pass into the belief in the existence of one or more gods. For savages would naturally attribute to spirits the same passions, the same love of vengeance or simplest form of justice, and the same affections which they themselves feel.

> ... a belief in all-pervading spiritual agencies seems to be universal; and apparently follows from a considerable advance in man's reason, and from a still greater advance in his faculties of imagination, curiosity and wonder.

And so there we have it. A Darwin leery of and with no personal use for organized religion, but with an understanding, respect, and firm recognition of the historic albeit mixed role of belief, i.e. spirituality, in evolution.

We have the Darwin who, were he here among us, in the face of environmental devastation, nuclear overkill, terrorism, and the horrifying rest of it, would surely tell a world wallowing in the nightmare of an evolutionary teen-age identity crisis to wake up and grow up in a hurry.

I shall argue that a predominant quality to be expected in a successful gene is ruthless selfishness...Much as we might wish to believe otherwise, universal love and the welfare of the species as a whole are concepts that simply do not make evolutionary sense.

Dawkins, *The Selfish Gene*

EIGHTEEN
THE RISE OF THE SUPER-NEOS

Other than the disastrous impact of the survival of the fittest mindset, the earlier rise of NeoDarwinian theory and science did much more to advance than to check in place or drive us backward in evolution. But then came the late 20th century emergence of the Super Neos. And this was a radically different matter.

The Moral Landscape by Sam Harris in 1994, with a reprint in 2011, and *The Better Angels of Our Nature* by Steven Pinker in 2012, signaled that a massive shift for the Super Neos back toward E.O.Wilson's original primacy for the moral sense might yet give the story a happy ending. But socially, politically, economically, and environmentally the damage of the Super Neos was devastating.

It will be said I'm exaggerating. But given the powerful impact of science on society—and of society on science—aren't we, in this and the next chapter, looking at what through incubation became the selfishess-*uber-alles* nightmare of the Donald Trump presidency?

The Super Neo's capture of the popular mind quietly began in

1975 with publication by Harvard University Press of the remarkable *Sociobiology: A New Synthesis*, by world-renowned Harvard entomologist E.O.Wilson.

In 1976 came the sure, swift touch of the Oxford University Press publication of *The Selfish Gene* by Oxford zoologist Richard Dawkins. Then in 1995 came publication by Simon and Schuster of *Darwin's Dangerous Idea*, by the Oxford expert on artificial intelligence, Daniel Dennett.

This trio of bright, gifted, and, in the case of Wilson and particularly Dawkins, prolific writers, set the pace for the avalanche of books that generated the profits for publishers ... that interacted with the ultra-conservative politics and economics of the Reagan-Bush I-Bush II years in America ... that spawned the vicious cycle that drove the massive lowering of expectations ... that accelerated the blind diminishing of species ... that set the stage for the shock of the Trump presidency and the worst of the turmoil thereafter.

Could this be, as will be charged, only the bias of "leftist elites," with no place in a supposedly properly impartial work of science?

As became blatant with Trump, this is a glimpse at a tactic used throughout the history of our species to convert the selfishness that Darwin called a "base principle" accounting for "the low morality of savages" to the gleeful celebration of selfishness as the sole master driver for human evolution.

E.O. Wilson

Critics quickly raised the question of Wilson's qualifications. How could a world authority on insect behavior and insect societies legitimately pontificate on human behavior and human societies? But in his passion for nature, the range of its subject creatures, and two crucial moves in line with all that had been pivotally ignored in

Darwin, Wilson's *Sociobiology* became a powerful historic pioneering.

In closing the gap between biological and cultural evolution that long blocked the advance of evolution theory, as we saw in chapter six, and in his rare new emphasis on the need for scientific attention to the question of morality and moral evolution, Wilson spoke to rising mass concerns.

Three quarters of the way through the 20th century, here was the scientific revival of far more than lip service to the *moral imperative,* which for the rest of Darwin was the primary driver for human evolution.

Here is Wilson reaching toward the stars in *The Diversity of Life,* Harvard University Press, 1992.

> There can be no purpose more inspiriting than to begin the age of restoration, receiving the wondrous diversity of life that still surrounds us."
>
> For what, in the final analysis, is morality but the command of conscience seasoned by a rational examination of the consequences. And what is a fundamental precept but one that serves all generations?
>
> An enduring environmental ethic will aim to preserve not only the health and freedom of our species, but access to the world in which the human spirit was born.

Wise and beautifully expressed, no doubt. But in the title for *Sociobiology*'s first chapter —The Morality of the Gene—out came the head of the worm in the apple. For after a magnificent attempt to set Super Neoism on the right course, Wilson also gave the Super Neos what they needed to collapse Darwin's vision of the cosmic regulation of the moral sense into nothing more than the "selfish gene" at work.

> "True selfishness...is the key to a more nearly perfect social contract," Wilson tells us in *On Human Nature.*

Further struggling to squeeze the vast range of Darwin's feeling for the moral sense into the hard and tight little ball of selfishness, Wilson tells us:

> "Human behavior...is the circuitous technique by which human genetic material has been and will be kept intact. Morality has no other demonstrable ultimate function."

In an astounding choice from the lexicon of pornography, he even moved on in *On Human Nature* to tell us that morality involves either the "hardcore altruism" of "kin selection" or the "soft-core altruism" of "reciprocal altruism."

Hamilton and Trivers

As if with the flash of a fan dance, Wilson thereby canonized the first saints for what, ironically, went on to in effect become a secular religion.

Based on studies of ants, bees, and wasps, in 1964 the noted biologist W.D.Hamilton locked in for Wilson and succeeding Super Neos ostensibly rock firm proof that at the core of altruism—that is, behavior helping others—lies the drive of selfishness favoring one's own relatives over non-gene-related strangers.

Then in 1971—based on an intimate study of vampire bats and naked mole rats—the noted biologist Robert Trivers surged beyond

Hamilton with ostensible further proof that in helping others, kin or not, *all of us are overwhelmingly governed by selfishness*.

This became the inflexible rule of "reciprocal altruism"— that we do good for others only because we expect this will motivate them to do good to us in return.

Thereby Hamilton and Travers became the patron saints for the blanket blessing of the work of Richard Dawkins, Daniel Dennett, and all Super Neo books thereafter.

Richard Dawkins

Beginning in the year following Wilson's *Sociobiology*, in 1976 with *The Selfish Gene*, Dawkins launched what book by book, by himself and then by others, steadily became gospel for the new faith.

As headed this chapter—and for a crucial reminder bears repeating:

> The argument of this book is that we, and all other animals, are machines created by our genes...I shall argue that a predominant quality to be expected in a successful gene is ruthless selfishness...Much as we might wish to believe otherwise, universal love and the welfare of the species as a whole are concepts that simply do not make evolutionary sense.

Also:

> We are survival machines—robot vehicles blindly programmed to preserve the selfish molecules known as genes. This is a truth which still fills me with astonishment. Though I have known it for years, I never seem to get fully

used to it. One of my hopes is that I may have some success in astonishing others."

Of "blind chance"—which both Neo-Darwinians and Dawkins changed from just plain unadorned "variation" for Darwin to "*random* variation" for his successors—we have this.

Darwin, as we've seen, had roared:

> The birth both of the species and of the individual are equally parts of that grand sequence of events that our minds refuse to accept as the result of *blind chance.*
>
> The understanding revolts at such a conclusion.

And over the floundering of all the intervening years, here was Dawkins' jaunty rebuttal.

> All appearances to the contrary, the only watchmaker in nature is the blind forces of physics, albeit deployed in a very special way. A true watchmaker has foresight: he designs his cogs, springs, and plans their interconnections, with a future purpose in his mind's eye. Natural selection, the blind, unconscious, automatic process which Darwin discovered, and which we now know is the explanation for the existence and apparently purposeful form of all life, has no purpose in mind.
>
> It has no mind and no mind's eye. It does not plan for the future. It has no vision, no foresight, no sight at all. If it can be said to play the role of watchmaker in nature, it is the blind watchmaker.

And:

> Nature is not cruel, only pitilessly indifferent. This is one of the hardest lessons for humans to learn. We cannot admit that things might be neither good nor evil, neither cruel nor kind, but simply callous—indifferent to all suffering, lacking all purpose.

And so on and on for the Doctrine of Dawkins ...

Daniel Dennett

And so we come to *Darwin's Dangerous Idea,* which on the book jacket Richard Dawkins tells us is "a surpassingly brilliant book" showing how "American intellectuals have been powerfully misled on evolutionary matters."

Widely popular, continuing to hammer across the gospel for Hamilton, Trivers, and selfishness, to many readers it seemed that here must be the absolute last word on Darwin and morality. For the book had not just one but actually two whole chapters on morality.

So we turn to the chapter "On the Origin of Morality" and what do we find? First comes Dennet's delight in the work of the ultra-rightist philosopher Thomas Hobbes—who tells us that by nature our species is designed to live lives that are "solitary, poor, nasty, brutish, and short." That society represents the "war of all against all." And that in this war "of every man against every man nothing can be unjust. The notions of right and wrong, justice and injustice, have there no place."

Hobbes, Dennett tells us, is the great founder of a scientific understanding of the origin and function of morality. Then come five pages of the vision of none other than Friedrich Nietzsche, the brilliant and amazingly gifted but pathologically sexist, racist, exultantly paranoid, and generally regressive philosopher who helped set the stage for the Nazis with his celebration of the crucial need for the rise of the "Superman."

This Superman, Nietzsche tells us, will be above such piddling concerns as the "slave morality" of Christianity and similar "weaklings."

Dennett assures us that Nietzsche has merely received a bad press from those who refuse to understand the true majesty of his thought. At the core for Nietzsche, he tells us, is Nietzche's "often outrageously misrepresented" concept of transvaluation as vital to the proper understanding of morality.

And what is this "transvaluation?" We are told by Nietzsche himself that it is the process whereby "the human soul in a higher sense acquire[d] depth and became evil"—the "two basic respects in which man has hitherto been superior to other beasts."

We are superior to other beasts through an evolution that has driven us to become evil?

This is what constitutes transvaluation to a supposedly higher level of morality?

Dennett assured us these were examples of Nietzsche's "terrific" Just So Stories. But the shocker, surely to at least some of his fellow Super Neos, must have been the fact that Dennett identified Hobbes and Nietzsche not only as our most useful basic moralists.

He says they're also the first two great sociobiologists.

> Scratch an 'altruist' and watch a 'hypocrite' bleed ... What passes for cooperation turns out to be a mixture of opportunism and exploitation.
>
> Ghiselin, *The Economy of Nature*

NINETEEN
SUPER NEOISM

Surely that was said to be funny, one thinks, as a rather poor bad joke. But let's have the full quote:

> Given a full chance to act in his own interests, nothing but expediency will restrain him from brutalizing, from maiming, from murdering—his brother, his mate, his parent, or his child. Scratch an 'altruist' and watch a 'hypocrite' bleed. No hint of genuine charity ameliorates our vision of society, once sentimentalism has been laid aside.
>
> What passes for cooperation turns out to be a mixture of opportunism and exploitation.

For many in the fields of social and natural science, and even more so in religion and the humanities, this was the wake up call.

It was perhaps not on the order of the biblical blowing of the horn of tocsin, but still universally jarring. For this quote, from *The Economy of Nature and the Evolution of Sex* by sociobiologist Michael

Ghiselin, came to be seen as a chilling glimpse into the heart and soul of the Super Neos.

This jolt helped speed the mutation of the Super Neos from phase one to phase two—that is, from sociobiology into the ostensibly more sensible and gentle gospel of evolutionary psychology.

Robert Wright

Alert to what looked like the prospect for a saleable new story within the hurly burly set off by E.O.Wilson with *Sociobiology,* in 1994 a bright enterprising journalist wrote what soon became the informal bible for evolutionary psychology.

Praised in *The New York Times* as "a feast of great thinking and writing about the most profound issues there are," and as the "most sophisticated in-depth exploration to date of the new Darwinian thinking" in *Publishers Weekly, The Moral Animal* by Robert Wright swiftly became enormously popular.

Here and there, on one level, it sought although fell short of living up to the billing. But on the level of our concern it again unfolded as with the flicker of another fan dance for the gospel of Super Neoism.

Again we were given the Super Neo bastion of Hamilton and Trivers. But now came a shock to rival Dennett's distortion. For Wright tells us that *The Descent of Man* is filled with Darwin's "doubts about the status of the moral sense."

He tells us that Darwin's belief that things have evolved for "the good of the group" is "mistaken." Instead, "Our ethereal intuitions about what's right and what's wrong are weapons designed for daily, hand-to-hand combat among individuals."

And why is this? Because "Darwin's moral sentiments" are "designed ultimately to serve selfishness."

Echoing Ghiselin (i.e., "scratch an 'altruist' and watch a 'hypocrite' bleed"), Wright finds "appeals for brotherly love are comparable to a politician's self-serving appeals to patriotism."

He tells us the "new paradigm is useful because it helps us see that the aura of rightness surrounding so many of our actions may be delusional; even when they feel right, they may do harm."

And what is to be our reward for taking this friendly advice to heart?

"If you understand the doctrine, buy the doctrine, and apply the doctrine, you will spend your life in deep suspicion of your motives."

This is how we should spend the rest of our lives?

Then comes an even more startling paragraph. One of the most important things the skilled professional writer develops is a keen, if not at times uncanny attunement to the mind of his or her targeted readership. So with shock we find Wright suddenly veering off the controlled page to take us to the core of the moral despair, the moral yearning, the moral rudderlessness, and eventually the moral avoidance of our age.

"So where does this leave us?" Wright asks. "Alone in a cold universe, without a moral gyroscope, without any chance of finding one, profoundly devoid of hope? Can morality have no meaning for the thinking person in a post-Darwinian world?"

Then, as if suddenly realizing these are dangerous waters, he deftly moved to reassure us.

"This," he told us, "is a deep and murky question that (readers may be relieved to hear) will not be rigorously addressed in this book."

Barkow, Cosmides, and Tooby

Again referencing the moral gospel of Hamilton and Trivers, in the next year, 1995, Oxford University Press published the formal bible for the new field.

Entitled *The Adapted Mind: Evolutionary Psychology and the Generation of Culture,* it was edited by anthropologist Jerome Barkow, an actual psychologist Leda Cosmides, and anthropologist John Tooby.

Among the book's eighteen chapters by others I found good ones designed to advance both theory and action in achieving evolution. But in light of mounting concern about what's being called the *singularity*—that is, whether the robotic spread of artificial intelligence will advance or eliminate our species, with crunch point purportedly by 2045—earlier we noted the fact that sociobiologist Daniel Dennett was considered an expert in this ominous field. Now, signaled by the editors in their introduction to *The Adapted Mind,* first came something with the unsettling flavor of a move in this direction.

> "Just as the fields of electrical and mechanical engineering summarize our knowledge of principles that govern the design of human-built machines, the field of evolutionary biology summarizes our knowledge of the engineering principles that govern the design of organisms, which can be thought of as machines built by the evolutionary process."

Was this a happy go lucky evocation of the artificially intelligent future? For a possible world of jobless humans scrambling for survival within the variations of a planetary takeover by robots?

Moving on from where sociobiology left off, along this line Barkow, Cosmides, and Tooby further signaled they had set out to write off a 300 year investment by philosophers, psychologists, and inspired educators in what had long been considered the single most powerful process advancing human evolution.

For they tell us that what all the deluded experts in these fields believed merely "reified this unknown functionality, imagining it to be a unitary process and called it '*learning*.'"

Learning not only remains "in genuine need of explanation," they wrote, but "will eventually disappear as cognitive psychologists and other researchers make progress in determining" what is really going on here.

They further predicted the same oblivion lies ahead for such concepts as "culture," "intelligence," and "rationality."

And what now of the exalted vision of the immense need for moral evolution, which E.O.Wilson had underlined and hailed in launching the Super Neo era?

I came to the conviction that Wilson was a great and admirable eminence in science, in short, in the vernacular, a "good guy." Opening up a feud that for a time titillated their followings, he put down Dawkins as a "mere journalist." As we saw in chapter eleven, he further departed the cohort he created to join the other Wilson, David Sloan, in fighting for community selection, i.e., "group selection."

Yet in Barkow, Cosmides, and Tooby, among hundreds of references and index entries for just about everything else, the single item referring to anything about moral sensitivity, moral development, or moral evolution is a passage of post-modernist academic mush under the revealing subhead "Conscience, Guilt, and Neurosis."

This was the brave new world to come when the delusions of all the great psychologists, sociologists, economists, and educators who came before the Super Neos were to be "replaced with knowledge"?

Darwin and the Fact of the Matter

The damage had already been done, but within the battle over evolution theory it's not far off the mark to say the Super Neos served to shoot the wounded and otherwise clean up the battlefield. Right on down their list for the "crimes" of the lost Darwin they went.

Morality is only the handy front for doing what best serves the interests of oneself and one's kin folk.

Love is the squishy word for something only meaningful in terms of "reproductive fitness" at the gene swapping level.

Sexual selection is a matter of battle and exclusive refinement at the gene swapping level.

Community selection, as we've seen, is a "dust bunny."

As for anything to do with self-organizing process or chaos theory, that was only something off in the fuzzy clutch of systems science, requiring a new set of references. And all this with the conviction one was guarding and advancing the one and only true, holy legacy of Darwin himself.

Surely by now it's evident this is not just more evidence of how the rest of Darwin was so effectively buried—and not far off to say, was murdered. Surely it's evident this is a matter enormously affecting the future of science, the future of our species, and ultimately the future of our planet.

Yes, we're motivated by selfishness—who in his right mind could deny it. This hasn't needed any untold long centuries of religion, or philosophy, or now science to prove what the least bit of human experience makes obvious.

The point—as Darwin repeatedly stated it—is the higher versus the lower motivation.

The Higher and the Lower Rules

Notwithstanding many sources of doubt, man can generally and readily distinguish between the higher and lower moral rules.

The higher are founded on the social instincts, and relate to the welfare of others. They are supported by the approbation of our fellow-men and by reason.

The lower rules, though some of them when implying self-sacrifice hardly deserve to be called lower, relate chiefly to self, and arise from public opinion, matured by experience and cultivation ...

In short, within the full sweep of Darwin's long ignored theory binding biological to cultural evolution lies both the scientific and social refutation of the idea that the evolution of our species is solely and wholly motivated by selfishness.

Indeed, how could anyone who claimed to be a scientist avoid, and for decades bog us down in ignorance of the size of the refutation that from all sides tells us what is real versus what is unreal, and what is true and what is false.

> May not the moral sense arise from ... our strong sexual, parental, and social instincts. May not this give "rise to 'do unto others as yourself'" and "'love thy neighbour as thyself." Therefore I say grant reason to any animal with social and sexual instincts and yet with passion he *must* have conscience.
>
> Darwin, *Early Notebooks*

TWENTY
SURF RIDING TOWARD THE SUPER SYNTHESIS

And so we are back to where we started.

Where did we go off track, and how can we get back on track in evolution?

We've come a long way from that sleepless night and discovery of those first clues that pointed to the burial of the rest of Darwin's theory. We've uncovered the shock of all that's come to light within this social scientific murder mystery we set out to solve. So what have we gained? Where and how did we go off track?

We've seen how burial of the rest of Darwin's theory led to the hole in modern mind that may swallow us in mass destruction.

Systems science tells us this is what you get if you obsess on the part and ignore the whole—and in terms of Darwin and Kant if you ignore the moral sense, and in terms of Jesus if you ignore love.

Given this dilemma, how could we—and indeed, within the blink in time still left us, *can* we get back on track?

I feel the evidence I uncover in this book shows why the future for our species and our planet depends on how soon the Gatekeepers and the Gatebreakers can get together.

In other words, it depends on how soon they can *work together* to heal the raw, open end of the wound that severed the first half from the second half of Darwin's theory, and thereby gain the progressive, moral and action oriented theory of guidance for our species that Darwin set out to build.

Trans or Super Synthesis?

Like a booster rocket for flight to our best destination in outer or inner space the scientists who transcended their squabbles, and joined Julian Huxley to build the great Neo-Darwinian synthesis, took us halfway there. Now what matters is how soon we can emulate them. What matters is how fast we can launch the slim second stage spaceship toward what we might, for the time being, call the transcendent synthesis, or Trans Synthesis, or in contrast to the Super Neo detour, the Super Synthesis.

Chapter by chapter we've seen how we seem to have been moving toward this goal at work among us. But the question I faced then—as you and all the rest of us face now—is will the progressive trend in the direction we've been going hold up?

What are the odds for our species versus the blind, the indifferent, and the mindless and the heartless.

The Case for Hope

In chapter eight we saw how for self-organizing Darwin and Wilber, Laszlo, and Eisler perceived it—and both the Neos and the Super Neos ignored it. We saw how for an important aspect of chaos theory Darwin and Wilber, Laszlo, and Eisler perceived it, and the Neos and the Super Neos ignored it. But what about the rest of the long buried or mangled factors for Darwin which, in chapters nine through seventeen, we've recovered?

What about *sex ... community .... love ...* and the central and centering drive of the *moral sense,* and open door for choice in the relation of science to *spirituality?*

For *sex*: Darwin, Wilber, and as a matter of gender Eisler elaborated it, Super Neos reduced it. For *community*: Darwin, Wilber, Laszlo, and Eisler treasured it, Super Neos trashed it. For *love*: Darwin, Wilber, and again particularly Eisler hailed it, Super Neos mocked it— e.g. "scratch an 'altruist' and watch a 'hypocrite' bleed." For the *moral sense*: Darwin and Wilber proclaimed it, Laszlo and Eisler embodied it. And with slam after slam of the boot kick of disdain, the Super Neos—failing to rise to E.O.Wilson's noble challenge—flattened the hapless moral sense to a squeak beneath the juggernaut roll of selfishness.

Wilber on Love and the Moral Sense

As for the problem that earlier stopped me, as others also came to feel about their differences, it melted within all that was rare and right in Wilber's vast spread of understanding. I came to know him, like him, and in wonderment see how out of the charisma of his "controversial" pioneering had emerged a remarkable contribution to the scientific, spiritual, and popular understanding of evolution.

He had gone beyond the vital but limited self-development guru stage to whole heartedly tackle the mighty power of the politics and economics of social change. Out of bright, gifted people animated by a passion for building the better world he had built what by any unbiased measure had become the most effective global organization to do this. All in all, I was struck by the feeling that here was a unique and remarkable embodiment of the Atman that Wilber had celebrated as a force for doing good in the world.

I decided to use his work for one last test of the factors of love and the moral sense in the Darwin-Gatebreaker connection.

For *love,* I found the tragedy that brought Wilber to life off the page: the five years he gave up practically everything else to nurse and care for his wife Treya, who was dying of cancer.

For Gatekeepers love tends to be seen as the squishy kind of thing one must avoid. But Wilber's book *Grace and Grit* is a deeply meaningful account of love not in lofty abstraction, but in both the earthy and transcendent grip of love's searing and soaring reality.

As he later wrote ...

> Love was here long before we were. It was here when this universe first exploded into existence. It was here when atoms first began to form molecules. It was here when those molecules first began to form cells. It's been here every step of the way—in fact, love is so fundamentally woven into the fabric of this universe that some even posit it as the fifth elementary force in the universe: the force of self-organization through self transcendence.

Then for me came the test of tests. Could I find in Wilber a connection to the fundamental insight that Immanuel Kant called the Categorical Imperative, which inspired the neglected Scotch Enlightenment, which in turn inspired Darwin, which filtered on down the line to development of my own moral transformation theory?

What I found was a dry and seemingly ice cold reference to "the moral intuition" as a matter of "greatest depth and greatest span."

Jabbed down in the margin alongside this passage in my copy of *A Brief History of Everything* is the fury of my first reaction. "Dreadful!" "Heartless!" "Offputting abstraction!" Like say an orange without juice.

But then on in *A Brief History of Everything* I found this concise restatement of the central factor for evolution—so tragically rare to find today, so hard for most of us to understand, so fundamental that philosophically Kant, and then scientifically Darwin, tried to firmly set in place.

> Q: You call this the Basic Moral Intuition.
>
> KW: Yes. The Basic Moral Intuition is "protect and promote the greatest depth for the greatest span." I believe that is the actual form of spiritual intuition.
>
> In other words, when we intuit Spirit, we are actually intuiting Spirit as it appears in all four quadrants ... as I and We and It ... Thus when I am intuiting Spirit clearly, I intuit its preciousness not only in myself, in my own depth, in my I-domain, but I equally intuit it in the domain of all other beings, who share Spirit with me ...
>
> And thus I wish to protect and promote that Spirit not just in me, but in all beings possessing that Spirit, and I am

> moved ... to *implement* this Spiritual unfolding in as many beings as possible: I intuit Spirit not only as I, and not only as We, but also as a drive to implement that realization as an Objective State of Affairs (It) in the world.

Unless you've basked in the exploration of the moral sense that still prevailed in the 19th century, this is likely hard to follow, but brilliant and powerfully to the point if you have—or clung to the old understanding despite all that has, like Sporting Life in Gershwin's magnificent opera Porgy and Bess, set out to water it down or destroy it.

Along with other important Gatebreaker contributions from Laszlo, Eisler, and others, I came to feel that Wilber's matrix could be vital for the Super Synthesis. The Gatekeepers might howl and scream and try to ignore it to the bitter end, but increasingly it seemed to fill a glaring hole in the Gatekeeper mind.

Wilber's matrix wasn't perfect, as ferocious critics swarmed to point out. Nothing ever is. But there was nothing else like it around and the more I looked at it the more I became fascinated by its possibilities.

We've glimpsed how century after century every effort to speed up the moral evolution of our species keeps getting bogged down for the lack of something that could bring everybody together on the same page.

Now here, it seemed to me, was a useful map to the territory that the cataclysm of disaster barreling down on us might at last force both sides to agree on before it's too late,

A map of the territory within which one may track, or select, the choice points out of which the future is formed.

For what it is, and how it works, see A Brief Guide to Wilber's Quadrants in Resources and Reflections ending this book.

But now rises the question that age after age has voiced the despair of our species. Could the need for a Super Synthesis to establish a moral action oriented theory of guidance possibly go anywhere?

How could this, or any other Gatebreaker advance, prevail against the dead, dragging weight of the paradigm, politics, economics, history and global drive of the survival of the fittest, selfish gene, and winner versos loser mindset stacked to attack even the slightest threat to the Sacred Status Quo?

As if guided by some planetary urge to make up for lost time, what's at stake for science, and for every aspect of our lives and future, can be seen in the juxtaposition of two radically different events in 2008.

The Keeper versus Breaker Reality

Over three days in July, 2008, sixteen highly credentialed biologists met at the Konrad Lorenz Institute, in Altenberg, Austria, for a symposium. It was impressively titled Toward an Extended Evolutionary Synthesis. It was heralded in advance as the bold move that "could turn into a major stepping stone for the entire field of evolutionary biology." Its purpose, as bluntly summarized by the noted science writer covering the event, Suzan Mazur, was "to begin sorting out the mess."

Rethinking evolution, Mazur wrote of the convictions of event organizers, was to be "pushed to the political front burner in hopes that 'survival of the fittest' ideology can be replaced with a more humane explanation for our existence and stave off further wars, economic crises, and destruction of the Earth."

"At a time that calls for scientific vision," she wrote, "scientific inquiry's been hijacked by an industry of greed, with evolution books hyped like snake oil at a carnival."

Out of that meeting, and a wide range polling of other leading biologists afterward, most impressive was the lack of things that even after 100 years these aroused and deeply concerned Gatekeepers could agree on.

They agreed the disastrous survival of the fittest mindset must be replaced. They agreed the new focus on self-organizing processes is a basic advance. But of the long ignored rest of all that Darwin actually wrote to complete his theory there's not a word.

Nothing about the astounding number of times Darwin wrote of the importance of love and the moral sense. Nothing on the jolt of surprise emerging from Darwin's move from the well-known biology of *Origin of Species* into the long ignored social and systems science of *The Descent of Man*.

Under the auspices of the prestigious Royal Society, a group of 300 contentious biologists met again in London in 2016, with more of the same purpose, ballyhoo, and lamentable lack of results.

But in sharp contrast to both of these globally heralded events, just one month later than the Altenberg symposium, in August 2008, 500 scientists and concerned others from 30 countries had gathered in the John F.Kennedy University in Pleasant Hill, California, for over 100 talks, panels, and poster presentations for the first biennial integral theory conference.

To underline the acid test for science in the real world the full title was Integral Theory in Action: Serving Self, Others, and Kosmos. Motivated by "a desire for a vision, a synthesis, that encompasses and frames the avalanche of information now upon us," the avowed goal was the "global framework"for "an Integral Age" in which "every single aspect of our work, play, education, medicine, economics, and self understanding becomes profoundly transformed."

Two events within two weeks of one another. On one side the quandary of the traditional Gatekeepers of evolution theory, who, with the real world breaking apart and calling for action all around us, are bogged down in a seemingly endless battle over woeful fragments of theory and territory.

On the other side, the free wheeling visions of a rapidly expanding band of Gatebreakers with heady claims for wedding the fragments into the expanding embrace of the greater theory.

Isn't this powerful further evidence of why the Gatekeepers and Gatebreakers—as well as the non-spiritual and the spiritual—must come together and work together to build what our species so urgently needs?

But will they?

Or as has been the history of more than a century for our species, will the challenge this recovery of the rest of Darwin raises be crowded out, shoved aside, and again fade leaving only a lone scattering of clues?

TWENTY ONE
THE DESTINATION OF SPECIES

To complete the cycle from origin to destination of species, Darwin turned to the question of what lies ahead for us three times in *The Descent of Man*. Long well known and frequently quoted is this first brief look ahead.

> Man may be excused for feeling some pride at having risen, though not through his own exertions, to the very summit of the organic scale; and the fact of his having thus risen, instead of having been aboriginally placed there, may give him hope for a still higher destiny in the distant future.

Then out of the optimism of his age he adds this vision—so haunting now, shattered in our age.

> Looking to future generations, there is no cause to fear that the social instincts will grow weaker, and we may expect that virtuous habits will grow stronger, becoming perhaps fixed by inheritance. In this case the struggle between our higher and lower impulses will be less severe, and virtue will be triumphant.

Not wanting to be accused of "evangelizing," elsewhere Darwin tells us he has purposely held himself in. But now he simply must give way to the excitement that first grips him and then his age—the new grandeur of the idea of evolution and evidence for the core driving idea of progress.

> The inhabitants of each successive period in the world's history have beaten their predecessors in the race for life, and are, insofar, higher in the scale of nature; and this may account for that vague yet ill-defined sentiment, felt by many palaeontologists, that organisation on the whole has progressed.
>
> As all the living forms of life are the lineal descendants of those which lived long before the Silurian epoch, we may feel certain that the ordinary succession by generation has never once been broken, and that no cataclysm has desolated the whole world. Hence we may look with some confidence to a secure future of equally inappreciable length. And as natural selection works solely by and for the good of each being, all corporeal and mental endowments will tend to progress towards perfection.

Still widely reverberating, as an argument for the Neos and a fervent subtext for the Super Neos, is the historic declaration of Jacques Monod, existentialist and post-modernist hero of the French resistance movement during World War II, who became a Nobel prize-winning molecular biologist.

Man at last knows that he is alone in the unfeeling

immensity of the universe, out of which he emerged only by chance. His destiny is nowhere spelled out, nor is his duty.

The kingdom above or the darkness below; it is for him to choose.

To this bleak view Monod added this punch to what became core doctrine for the Super Neos and Super Neoism.

"Anything can be reduced to simple, obvious, mechanical interactions. The cell is a machine; the animal is a machine; man is a machine," Monod wrote in 1970. Variation, he said in 1972, was an "accident" drawn "from the realm of pure chance."

Again we're looking at the vital ingredient dropped from the rest of Darwin. For as we've seen, roaring like a rocket out of The Descent of Man and his final years, is this scientific and spiritual rejoiner.

> The birth both of the species and of the individual are equally parts of that grand sequence of events that our minds refuse to accept as the result of blind chance.
>
> The understanding revolts at such a conclusion.

This was the flare in the night that over 100 years reaches out from the buried rest of Darwin to all those in our time who work to build the better world.

This was and is the stance, the drive, the path, the stream of recognition and the ardent cause enlisting all those in our time still driven by the vision of a great destination of species.

The Passing of the Torch

This was the vision that seized and for nearly two decades drove our little group after the meeting in Budapest. This was the vision that gave us hope we could help end the threat of nuclear annihilation hanging over us.

Every year the great cities of Europe had competed for the honor of being designated Europian Cultural Capitol for that year. Our General Evolution Research Group symposia became the centerpiece for Florence's year of global fame.

There we had been at the launching point for the Renaissance with all its wonders now within a short walk in between our performances.

Michelangelo's incredible soaring seventeen foot high statue of David. The Uffizi Gallery, the world's oldest with works by da Vinci and Botteschelli. The splendor of the Pitti Palace with Titian, Rubens, and Raphael. The Ponte Vecchio spanning the Arno—the Renaissance first model for the upscale shopping mall, glittering with expensive shops and the exotic flow of lookers and shoppers. And over it all, wherever you walked, the soaring presence and slumbering magnificence of Brunelleschi's Duomo.

From physics and biology on up the line through the social sciences into systems science, the routine for each of us was to step up in turn to the lectern and present our prestigious papers on the potential of our particular field for personal and planetary evolution.

Next we became the leading attraction for Bologna's celebration of the 900th anniversary of the founding of the world's oldest university. More of the same for Rome, Vienna, Berlin, Turku in

Finland, Sardinia, and last the two symposia I organized for the World Congress of the Systems Sciences in Toronto in 2000, from which we issued the Toronto Manifesto to serve as an elegy, celebration, and call to arms.

To try to reach and build the support our mission needed our papers had been published in issues of our flagship journal, *World Futures: The Journal of General Evolution,* and other influential publications. They'd been recycled in a series of books published by Feltrinelli in Italian and worldwide in English by Taylor and Francis, giant publisher of more than 1,000 journals and over 1,800 new books each year.

We'd had a colorful time of it. At one point GERG was officially headquartered in the presumably sumptuous manor of Fiona, Baroness Montagu of Beaulieu. Another time it was the palace of Prince Alfred of Liechtenstein—a particularly fetching fellow, nearly seven feet tall, with a Chinese girl friend literally half his size.

On the lighter side had been the delight in jokes, stories, laughter, the excitement of mobilized intelligence, and the sense of a brave band of brothers and a sister all in this together, which I've tried to bring to life. On the heavier side, however, was the ominous drum beat of the news.

The fear that grew not just in prospect, as with the nuclear nightmare, but in the steady chop of reality. The poisoning of land, air, water, and minds. The insanity of terrorism. The signs of a rebirth of the fascism and anti-Semitism that gave us Hitler and the Holocaust. The political shift from progression to regression. The escalating gap between rich and poor. The conspiracy of forces chopping away at all hope for a livable future, and everything else that became a grim reminder of our mission and responsibility.

Buoyed up and blithely confident of impact we drew comfort from knowing that our feelings were globally even passionately mirrored in thousands of independent scientists, scholars, and institutes concerned about the destination of our species. Here by the year 2000 I think pretty much all of us felt we had entered what increasingly looms as the cliff edge crunch point in human evolution. As had led me to actually read Darwin and find the clues to what had happened, it became more and more apparent that less and less was moving ahead either for the theory or the reality of evolution.

In comparison with the power of the global death grip of the pseudo-Darwinian mindset of "survival of the fittest," "the selfish gene," and then all too soon the disastrous "winner versus loser" mindset of the disastrous Trump presidency, was the long range prospect for our species and all we loved and valued hopeless? For by now the evidence from field after field of both natural and social science was practically screaming at us that this could be, indeed even likely be our fate.

During our years, Eric Chaison saw the reach of the Hubble Telescope into far outer space completed. Prigogine and Varela basked for a time in the fame of being self-organizing process theory founders. Ralph and Fred Abraham did the same for chaos theory. Karl Pribram's theory of the holonomic brain and much else widely took hold. As for Darwin's central concern with moral evolution, Wilber and Laszlo continued to embody it and I had fought uphill with twelve books, most still unpublished.

For me the most meaningful hope for a better future was signaled by a new classic study for its field. In *Macrohistory and Macrohistorians,* the editors chose the thinkers they considered the twenty most important historically—among them Persian philosopher Ibn Kaldun, both Karl Marx and Adam Smith, Arnold Toynbee, Pitirim Sorokin, Teilhard de Chardin, Oswald Spengler, Max Weber ... and the first woman and first living human to be admitted to this Generally Bearded Old Boys Club of Major Thinkers: Riane Eisler.

This and much more was all to the good, but considering all that our species is now up against, had the work of our group, and internationally thousands more like us, made no difference?

Like the tiny raft of Kontiki, or the tiny long boat of Leif Erickson, or the pilgrim's tiny Mayflower caught in the indifferent rage of the mightiest of storms, could the vast wrangling mass of our species be pointed toward the fulfillment of life on earth rather than its destruction?

Generation after generation, with the passing of the torch we older ones seek to arm fresh runners with the vision of a great destination of species. Century after century each new generation has pushed on to bring us to where we are today. But with all around us the fear spreading that ours may be at the end game place in evolution, belief in a better future is falling apart. Yet once again new generations must jump in to take up the torch from us and push ahead.

In this recovery of the rest of Darwin, haven't we gained powerful support for the better theory of evolution and the better world?

Can we now see, and understand, and wake to fight for the difference that a theory not just in our minds, but that could match the hope still in our heart and soul, can make?

A theory that for the first time in terms of both science and spirituality could serve as a guide to the better future that our species has struggled to gain for over 100,000 years?

Whoever and wherever you are, young, old, or in-between, take up this new sense of all that's built up over time within us and help put it to work for us.

With a new sense of who we can, should, and now *must* be, call for, insist on, relentlessly demand of science and spirituality, and if needed *fight for* and once gained put to work, the guide to the better future for which our time cries out.

To help arm, bless, and speed your voyage, here are four hard won perspectives on the destination of species—first the third and most complete of Darwin's views; then toward the same end, updates from Ken Wilber, Ervin Laszlo, and Riane Eisler.

Darwin on the Destination of Species

Typically Darwin packed his main glimpse into the future into a tight little paragraph that for over a century has been easy to skip past as a stray oddment of doubtful relevance.

I've opened it up to give the lines their innate bardic roll.

The social instincts acquired by us as by the lower animals for the good of the community from the first have given us the wish to aid our fellows, feelings of sympathy, and the desire to seek the approval and fear of the disapproval of others.

These impulses served in the early development of our species to provide a rude rule of right and wrong. But as our species has gradually advanced in intellectual power, and thus has been enabled to trace the remote consequences of our actions;

and has thereby acquired sufficient knowledge to reject baneful customs and superstitions;

and as we have regarded more and more not only the welfare but the happiness of our fellow beings;

and as then from habit, following on beneficial experience—extending to people of all races, to the imbecile, to the maimed, to other members of society, and finally to the lower animals,

so has the standard of our morality risen higher and higher.

Ken Wilber on the Destination of Species

Are the mystics and sages insane? Because they all tell variations on the same story, don't they? The story of awakening one morning and discovering you are one with the All, in a timeless and eternal and infinite fashion. Yes, maybe they are crazy, these divine fools. Maybe they are mumbling idiots in the face of the Abyss. Maybe they need a nice, understanding therapist.

Yes, I'm sure that would help. But then, I wonder. Maybe the evolutionary sequence really is from matter to body to mind to soul to spirit, each transcending and including, each with a greater depth and greater consciousness and wider embrace. And in the highest reaches of evolution, maybe, just maybe, an individual's consciousness does indeed touch infinity—a total embrace of the entire Kosmos—a Cosmic consciousness that is Spirit awakened to its own true nature. It's at least plausible. And tell me: is that story, sung by mystics and sages the world over, any crazier than the scientific materialism story, which is that the entire sequence is a tale told by an idiot, full of sound and fury, signifying absolutely nothing?

Listen very carefully: just which of those two stories actually sounds totally insane?

Ervin Laszlo on the Destination of Species

In the embracing vision that is now emerging everything that has evolved in the universe, Mozart and Einstein, you and me, the greatest of galaxies and the humblest of insects, is the result of a stupendous process of open-ended yet non-random self-creation.

Nothing that has ever evolved exists separately from all the rest: all things are connected, all are part of an organic totality.

We experience this field as a seamless whole composed of its parts. More than that, it is a whole in which all parts are constantly in touch with each other. There is constant and intimate contact among the things that co-exist and co-evolve in the universe, a sharing of bonds and messages that makes reality into a stupendous network of interaction and communication.

At a time when we and our societies transit into an interacting and interdependent web of technology, finance, production, consumption, and even leisure and culture, it is vital that our consciousness be infused with this new vision, rather than the old.

It is the kind of insight that could re-establish harmony and balance in a world of vulnerable interdependence and growing chaos.

Riane Eisler on the Destination of Species

Why do we hunt and persecute each other? Why is our world so full of man's infamous inhumanity to man—and to woman? How can human beings be so brutal to their own kind? What is it that chronically tilts us toward cruelty rather than kindness, toward war rather than peace, toward destruction rather than actualisation?

This is how she opens The Chalice and the Blade. And coining the word gylany *to describe it ... (*gy *from* gyne *for woman and* an *from* andros *for man, with* l *for the link between them) ... this is the future she sees still open to our choice if we speed the shift from a domination to a partnership way of life.*

For above all, this gylanic world will be a world where the minds of children—both girls and boys—will no longer be fettered ... our drive for justice, equality, and freedom, our thirst for knowledge and spiritual illumination, and our yearning for love and beauty will at last be freed.

And after the bloody detour of androcratic history, both women and men will at last find out what being human can mean.

EPILOGUE
WHO DID IT, WHY, AND WHAT NOW?

So once again:

Where did we go off track, and how can we get back on track in evolution?

The quick answer, I'm tempted to say, is this:

Isn't this what you get if you chop a good theory of evolution in half and try to run the world with the worst half for that purpose?

After all the years of this investigation of what we've been told was all of Darwin —and *all* we could expect of *ourselves*—what have we uncovered?

In jarring contrast to the popularity of what became the Darwin of "survival of the fittest,""selfish genes" —and the rampage of "winners versus losers"— let's take another look at [our] recovery of the five factors to *"speed the evolution of our species."*

First, prefiguring Freud, two chapters focused on Darwin's long-ignored higher-order understanding of ***sex***. Then, prefiguring work comparatively recent in science, came his long-buried exploration of the basic drive and fundamental nurturance of ***love***. Same for the global bond of ***community***— or "***mutual aid***," as it was called it in Darwin's time. And of greatest importance—long recognized by everyone from Jesus in religion to Immanuel Kant in philosophy—was how Darwin pounded away insisting on the primary drive of ***the moral sense in human evolution***.

Here too we saw—and many will feel—the shock of Darwin's long-ignored case for ***spirituality*** and the place and function of the *positive* basic teachings of ***religion*** in evolution. Still further came the surprise of how, in what he wrote of "the morality of women," Darwin emerges as a cautious forerunner of male support for the women's movement.

All in all, stage by stage, we've seen how, beginning in the 19th century then spanning the 20th into the 21st century, the rest of Darwin was wiped off the slate of history. But now—and for my part in a shake of the fist—comes the question of who did it, how, why, and what now.

Although I had glimpsed it earlier, it was slow to sink in. Only at the last— in a sudden flash of light in many directions, as this book went into, and had to be inserted in, the galley proof stage—did I see it clearly.

Here it was and I was, two, going on three decades of the work that led to this book, which had been reported in journals and other books and hailed by the scientists and others who were following it as a work in progress. Yet only now did there fully burst into mind the most startling and meaningful fact that *it wasn't really new.*

Sex, love, community, the moral sense, spirituality, and the drive for freedom and equality—experienced by our species and written about for now thousands of years, the only thing new about it is that it is what almost all of Darwin's successors dropped from what Darwin originally believed, said, and wrote for posterity.

Here were all these years of saying and proving the same thing over and over again, and still it could be wiped from the the guiding mind of our species.

What does this tell us of how science can be taken over and warped and twisted not only to slow the evolution of the best among us, but to tighten the death grip of the worst among us?

What does this tell us about a structure for our worldwide society based on a corruption of the first half and burial of the completion of his theory?

What does this tell us about what has and is being done to politics, economics, morals, spirituality, and just about everything else that shapes us for better or worse?

So much to gain, and saver, and know and put to use; so little time now left for our species to try to unpack within a single century all that would normally require another thousand years.

With all that is left in me I urge you to further probe, ponder, tell about, teach, stand up for, and with everything in you, if necessary, *fight* for what this book—and all the bright, brave people in it—show you we need to do to get to where we need to go.

REFLECTIONS AND RESOURCES

JOIN THE NEW X CLUB FOR THE NEW DARWIN

Will the Darwin of "survival of the fittest" and "selfish genes" or the Darwin of love and the moral sense prevail?

This is a classic case of the eternal clash between an Old and a New Paradigm. It is, however, much more than that. For the burial of the rest of Darwin and what we do about it—or don't do about it—could be the end game test for our species.

Well known historically is the famous X Club originally formed by Darwin's famous "bulldog" Thomas Huxley. Including Hooker, Henslow, and other famous scientists in that time, it was X Club support for Darwin—then mainly known as only a popular travel book writer—that gained the toe hold for his climb up the mountain to establish his great heresy against fierce opposition within the science as well as the religion of his time.

Formation of the New X Club for the New Darwin can do the same for us. Recovery of the buried Darwin of love and the moral sense as prime drivers of evolution again faces fierce opposition from everything with an investment in the environmentally, politically, economically, and spiritually destructive "Darwinian" survival of the fittest/selfish genes mindset.

The powerful *pre-publication reviews* with which I opened this book show why the formation of a New X Club for the New Darwin could gain the global membership that can give us a fighting chance to break through to better days.

Add your name to the record of those who want to help the Darwin of love and the moral sense prevail. Join your mind, heart and soul mates in writing a new history of hope for our children, our species, and our battered planet.

The New X Club for the New Darwin

Join to build the better world !

My name ______________________________

State or Country ______________________________

Email address ______________________________

Comment?

To join send to

darwinxclub@gmail.com

A BRIEF GUIDE TO WILBER'S QUADRANTS

A two page visual spread for the Quadrants follows this brief guide to their structure and meaning.

To gain an understanding of what may otherwise evade you first note the big X in the middle. This provides the anchoring logic of X for the hypothetical Big Bang for cosmic evolution that sets everything in motion. Then note how out from this cross point in the middle the action explodes in four directions.

Along the top right thrust for the X we see the stage by stage development of everything in our world from the atom to consciousness.

Along the top left thrust you see the stage by stage growth of everything within us and all life forms, from emotion to the consciousness, which gives us the capacity to make sense of our world.

Bottom left is the evolutionary thrust for all life forms' cultural development. Bottom right is the thrust of evolution from galaxies and planets to tribes and all other forms of social organization up to where we are today.

Thus out from the mid page starting point in the Big Bang you can get a useful sense of how much of everything we know of the multi-billion year development of all that became life on earth, including ourselves, unfolds.

There is more to it—so much it may look too complex, maybe something to flip past and come back to later. In one way or another, however, if you persist it is as though what is here on a flat page becomes a hologram.

That is, as though the flat page becomes a multiperspectival floating image one can turn about in mind to see how, within a good chunk of what was, is, and should be, everything interconnects.

Wilber and dedicated Wilberians working with him— as of this writing, e.g., Jeff Saltzman, Sean Esbjorn Hargens, Steve Mackintosh, Carter Phipps, Terry Patten, Michael LaGattuta, and many others globally—have successfully put the Quadrants to use in a wide variety of personal, social, political, and economic as well as spiritual problem areas. (See Esbjorn Hargens' *Integral Theory in Action,* or currently Jeff Saltzman's Daily Evolver internet series).

Shown here first is the whole matrix sqashed onto a single page.

So you can see and follow otherwise illegible details, next come separate close ups for both halves.

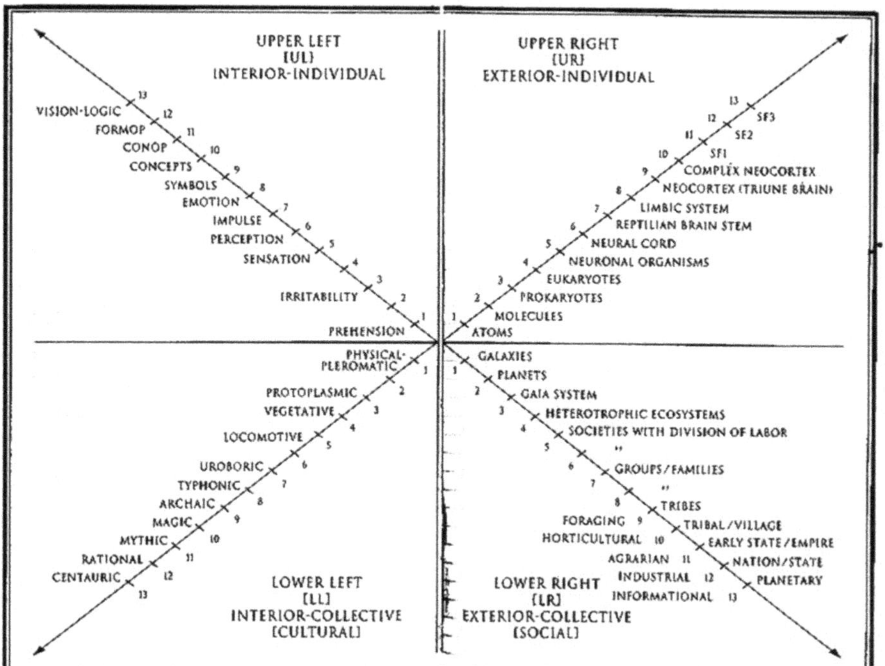

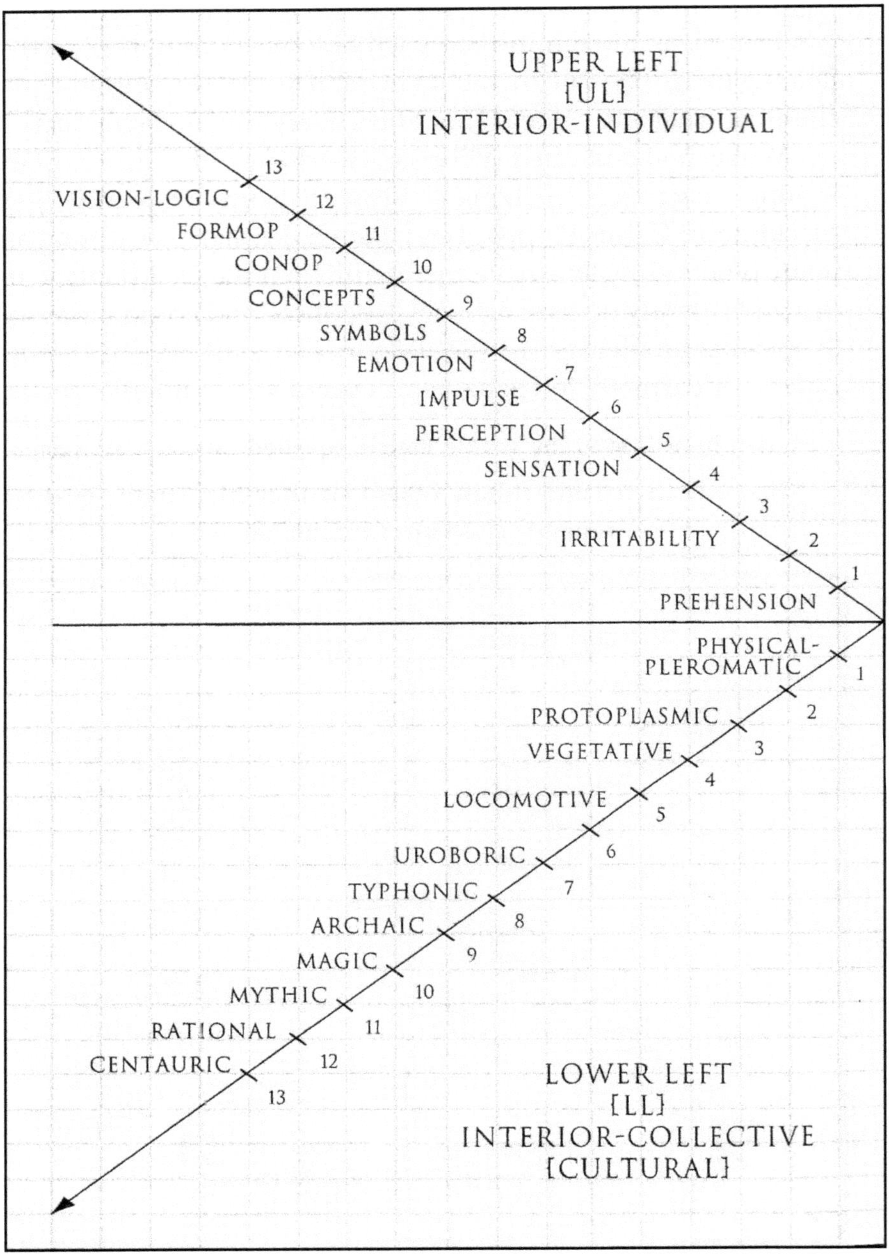
UPPER LEFT
[UL]
INTERIOR-INDIVIDUAL
13
VISION-LOGIC
12
FORMOP
11
CONOP
10
CONCEPTS
9
SYMBOLS
8
EMOTION
7
IMPULSE
6
PERCEPTION
5
SENSATION
4
3
IRRITABILITY
2
1
PREHENSION
PHYSICAL-
PLEROMATIC
1
2
PROTOPLASMIC
3
VEGETATIVE
4
LOCOMOTIVE
5
UROBORIC
6
TYPHONIC
7
ARCHAIC
8
MAGIC
9
MYTHIC
10
RATIONAL
11
CENTAURIC
12
13
LOWER LEFT
[LL]
INTERIOR-COLLECTIVE
[CULTURAL]

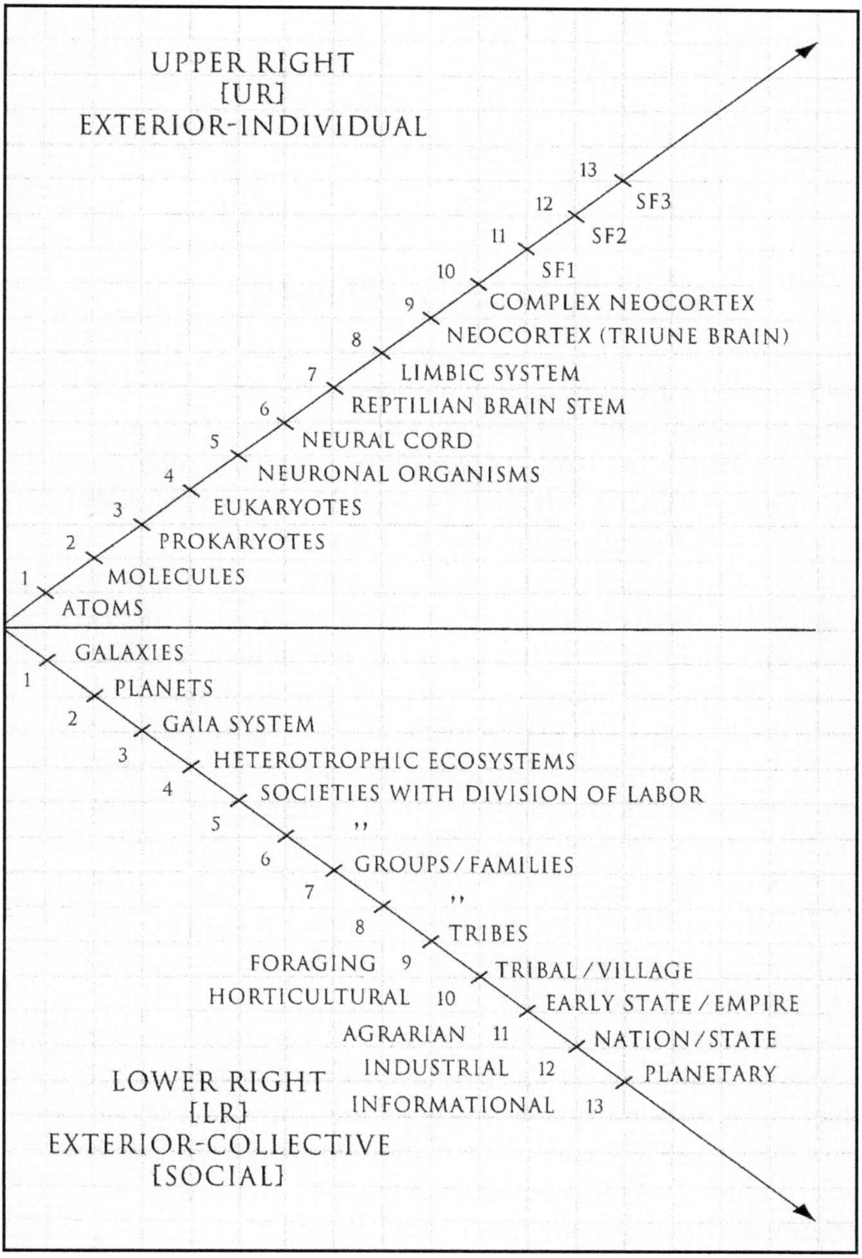
UPPER RIGHT
[UR]
EXTERIOR-INDIVIDUAL
13
SF3
12
SF2
11
SF1
10
COMPLEX NEOCORTEX
9
NEOCORTEX (TRIUNE BRAIN)
8
LIMBIC SYSTEM
7
REPTILIAN BRAIN STEM
6
NEURAL CORD
5
NEURONAL ORGANISMS
4
EUKARYOTES
3
PROKARYOTES
2
MOLECULES
1
ATOMS
GALAXIES
1
PLANETS
2
GAIA SYSTEM
3
HETEROTROPHIC ECOSYSTEMS
4
SOCIETIES WITH DIVISION OF LABOR
5
,,
6
GROUPS/FAMILIES
7
,,
8
TRIBES
FORAGING 9
TRIBAL/VILLAGE
HORTICULTURAL 10
EARLY STATE/EMPIRE
AGRARIAN 11
NATION/STATE
INDUSTRIAL 12
PLANETARY
INFORMATIONAL 13
LOWER RIGHT
[LR]
EXTERIOR-COLLECTIVE
[SOCIAL]

ABOUT THE AUTHOR

David Loye is a psychologist, evolutionary systems scientist, and author of thirty books including the national award winning *The Healing of a Nation* and *Darwin's Lost Theory*. A former member of the Princeton and UCLA School of Medicine faculties, professor in the research series for the Neuropsychiatric Institute, he is best known for his fight to update and expand evolution theory following his discovery of the long buried rest of Darwin's theory emphasing love and the moral sense, *not* survival of the fittest and selfish genes, as primary higher order drivers of evolution.

During the twenty years of research that led to his recovery of the rest of Darwin's theory, Dr.Loye has been involved with scientists in many fields and countries in developing the cutting edge fields of chaos, complexity, integrative, and evolutionary systems theory. He is a co-founder of two international organizations for updating and expanding evolution studies: The General Evolution Research Group, and The Society for Chaos Theory in Psychology and the Life Sciences.

He is also the co-founder, with the cultural evolution theorist and well-known author of *The Chalice and the Blade,* Riane Eisler, of The Center for Partnership Studies (www.partnershipway.org); and founder of The Darwin Project (www.thedarwinproject.com), with a Council of more than 50 progressive American, European, and Asian scientists, educators, and media activists.

Currently, at 92, he is trying to live long enough to complete the four books in which he develops his multidimensional Moral Transformation Theory: *Rescovering Goodness, Redefining Evil, Redefining Morality*, and *Transformation.*

A World War II veteran married to a Holocaust survivor, he lives with his wife, Riane Eisler, in California.

ACKNOWLEDGMENTS

To think back upon all who helped you during the years it took to get to this point is a moving and humbling experience. Where do you begin, where do you end?

I think of all the fascinating people I came to know and learn from. Beyond those I've briefly written of in this book was the amazing scientific diversity and global spread of others within both our General Evolution Research Group and the Council of the Darwin Project.

Ranging from biologists and physicists to psychologists, sociologists, philosophers, historians, educators, and political, management and systems scientists ... from Great Brittain, Belgium, France, Germany, Sweden, Hungary, Switzerland, Finland, Italy, and Russia in Europe and China, Sri Lanka, and Australia in Asia ... here are both those identified in the book and the rest of the members of both groups to whom I'm indebted.

For the General Evolution Research Group

Ervin Laszlo, Peter Allen, Robert Artigiani, Ralph Abraham, Singa Sandelin Benko, Kenneth Busch, Gianluca Bocchi, Thomas Bernold, Raymond Bradley, Bela A.Banathy, Bela H. Banathy, Alexander Christakis, Allan Combs, Miriam Campanella, Mauro Ceruti, Eric Chaisson, John Corliss, Mihaly Csikszentmihalyi, Vilmos Csanyi, Duane Elgin, Riane Eisler, Sally Goerner, Attila Grandpierre, Susantha Goonatillake, Mae-Wan Ho, John Hisnanick,

Min Jianin, Stanley Krippner, Jurgen Kurths, Gyorgy Kampis, David Loye, Alexander Laszlo, Kathia Laszlo, Eduard Makarjan, Ignacio Masulli, Pentti Malaska, Alfonso Montuori, Mika Pantzar, Ilya Prigogine, Karl Pribram, Gerlind Rurik, Maria Sagi, Peter Saunders, Stanley Salthe, Jonathan Schull, Rudolf Treumann, and Francisco Varela

For the Darwin Project Council

Marcus Anthony, Angeles Arrien, Ralph Abraham, Bela H. Banathy, Kenneth Busch, Richard Bird, Howard Bloom, Raymond Bradley, Alexander Christakis, Allan Combs, Gerald Cory, Jr., Mihaly Csikszentmihalyi, Riane Eisler, Duane Elgin, Sally Goerner, Rod Gorney, Thom Hartmann, Hazel Henderson, Mae-Wan Ho, Barbara Marx Hubbard, Sohail Inayatullah, Min Jiayin, Jeffrey Kane, Helena Knyazeva, Stanley Krippner, Hans Kung, Ervin Laszlo, Daniel Levine, Bill Levis, David Loye, Paul D. MacLean, Peter Meyer-Dohm, Ron Miller, Alfonso Montuori, Nel Noddings, Bruce Novak, John O'Manique, Barclay Palmer, Sister Ruthmary Powers, Karl Pribram, Raffi, Robert J. Richards, Ruth Richards, John Robbins, Nancy Roberts, Frank Ryan, M.D, Stanley Salthe, David Scott, Tim Seldin, Christine Sleeter, Joseph Subbiondo, Brian Swimme, and Michael Toms.

Significant Others

Besides cherished and meaningful members of my two families, I am further indebted to, and want to thank those I came to know through the invaluable experience of my two year membership in the Unitarian Universalist Church of Monterey Peninsula; my Thursdays with the explorations of the Elders Group; my Il Fornao

Friday buddies, Lou, Kim, and Brian; my fortunate involvement in the Integral Theory community inspired by Ken Wilber's works and impressive team, Jeff, Steve, Terry, Carter; and Michael—who did a magnificent job identifying Darwin quote sources; and the abiding friendship and continuing aid of the incomparable Opera Star and publisher David Gordon.

Through it all, always, in both former, present, and I hope future lifetimes, the incomparable love of my brilliant and beautiful wife and partner Riane.

BIBLIOGRAPHY

A Special Readers Guide to the three considerably different editions of Darwin's The Descent of Man listed in this bibliography.

The Princeton University Press edition is the widely entrenched but cumbersome two volume 1871/1971 first edition immensely improved in later single volume editions.

The best of these is Darwin's 1879 edition, which Penguin published with exemplary editing and commentary by leading Darwin biographers **Adrian Desmond and J.R.Moore** in 2004.

Penguin followed, in 2007, with a special edition with an editing of the Desmond/Moore 1879 text, by the noted science writer **Carl Zimmer** into a very useful re-ordering of the often confusing sprawl of *Descent* into separate categories, e.g., for hominid evolution, mental powers, morality, civilization, race, and sexual selection in animals and humans.

This "bare bones" bibliogaphy for ***Rediscovering Darwin*** will be expanded with further text, notes, references, and index in an companion edition to be published when its author completes this vital but woefully demanding final task.

Abraham, F., Abraham, *R.,* and Shaw, C. *A Visual Introduction to Dynamical Systems Theory for Psychology.* Aerial Press, 1990.

Abraham, R., and Shaw, C. *Dynamics: The Geometry of Behavior.* Addison-Wesley, 1992.

Abraham, R. *Chaos, Gaia, and Eros*. HarperSanFrancisco, 1995.

Adorno, T.W., Frenkel-Brunswick, E., Levinson, D.J., and Sanford, R. N. *The Authoritarian Personality.* Harper, 1950.

Allport, G. *Becoming.* Yale University Press, 1955.

Barkow, J., Cosmides, L., and Tooby, J. *The Adapted Mind.* Oxford University Press, 1992.

Bausch, K. *The Emerging Consensus in Social Systems Theory*. Plenum, 2000.

Bausch, K., and Christakis, A. Technology to Liberate Rather Than Imprison Consciousness. In Loye, D., Ed., *The Great Adventure: Toward a Fully Human Theory of Evolution.* SUNY Press, 2004.

Bellah, R., Madsen, R., Sullivan, W., Swidler, A., and Tipton, S. *The Good Society*. Knopf, 1991.

Bradley, R.T. *Charisma and Social Structure: A study of love and power, wholeness and transformation*. Paragon House, 1987.

Bradley, R.T. Love, Power, Mind, Brain, and Agency. In Loye, D. (ed.) *The Great Adventure: Toward a Fully Human Theory of Evolution*. SUNY Press, 2004.

Bradley, R.T. Psychophysiology of Intuition: A quantum-holographic theory of nonlocal communication. *World Futures: The Journal of General Evolution*, Vol. 63(2), 61-97, 2007.

Chaisson, E. *The Life Era*. Atlantic Monthly Press, 1987.

Chaisson, E. *Cosmic Evolution*. Harvard University Press, 2000.

Combs, A. *The Radiance of Being*. Paragon House, 2002.

Combs, A. *Consciousness Explained Better*. Paragon House, 2009.

Csanyi, V. *Evolutionary Systems and Society*. Duke University Press, 1989.

Csikszentmihalyi, M. *The Evolving Self*. Harper Perennial, 1994.

Csikszentmihalyi, M. *Creativity : Flow and the Psychology of Discovery and Invention*. Harper Perennial, 1996.

Darwin, C. *The Origin of Species* (150th Anniversary Edition, with introduction by Julian Huxley). Signet, 2003.

Darwin, C.. *The Descent of Man* (cumbersome first edition). Princeton University Press,1871/1981

Darwin, C. *The Descent of Man* (with updating of 1879 best edition and introduction by Darwin biographers Adrian Desmond and J.R.Moore). Penguin, 2004.

Darwin, C. *The Descent of Man* (with useful re-ordering of sprawling text by noted science writer Cal Zimmer). Penguin, 2007.

Darwin, C. *Autobiography*. Norton,1887/1993.

Darwin, C. *The Expression of the Emotions in Man and Animals*. University of Chicago Press, 1965

Dawkins, R. *The Selfish Gene*. Oxford University Press, 1976.

Dawkins, R. *The Blind Watchmaker*. Norton, 1987.

de Chardin, T. *The Phenomonon of Man*. Fontana, 1955.

de Chardin, T. *On Love and Happiness*. Harper Collins, 1984.

Dennett, D. *Darwin's Dangerous Idea*. Simon & Schuster, 1995.

De Waal, F. *Goodnatured*. Harvard University Press, 1996.

Desmond, A., and Moore, J. *Darwin: The Life of a Tormented Evolutionist*. (often rated best biography) Penguin, 1991.

Dobzhansky, T. *Mankind Evolving*. Yale University Press,1987.

Eisler, R. Action Research and Human Evolution: David Loye's Lifelong Exploration of Moral Sensitivity. *World Futures: The Journal of General Evolution* (1997): 49, 1-2,

89-101.
Eisler, R. *The Chalice and the Blade*. Harper & Row, 1987.
Eisler, R. *Sacred Pleasure.* HarperSanFrancisco, 1995.
Eisler, R. *Tomorrow's Children.* Westview Press, 2000.
Eisler, R. *The Power of Partnership.* New World Library, 2001.
Eisler, R. *The Real Wealth of Nations: Creating a Caring Economics*. Berrett-Koehler, 2008.
Esbjörn-Hargens S. and Wilber, K. *Integral Theory in Action.* SUNY Press, 2010.

Fox, M. *Original Blessing.* Tarcher, 2000..
Fromm, E. *Man for Himself: An Inquiry into the Psychology of Ethics.* Holt, Rinehart, and Winston, 1947.

Galtung, J., and Inayatullah, S., Eds. *Macrohistory and Macrohistorians*. Westport, CT: Praeger, 1998.
Ghiselin, M.T. *The Economy of Nature and the Evolution of Sex*. University of California Press, 1974.
Goerner, S. *After the Clockwork Universe*. Floris Books,1999.
Goerner, S. *Chaos and the Evolving Ecological Universe.* Routledge, 1994.
Goldie, P. *Darwin 2nd Edition.* A CD-ROM containing major books and papers by Darwin produced by Lightbinders, Inc., 1997.
Gorney, R. *The Human Agenda.* Simon and Schuster, 1972.
Gould, S.J. *Ever Since Darwin.* Norton, 1980.
Gould, S.J. *The Structure of Evolution Theory.* Harvard University Press, 2002.
Greene, J. *The Death of Adam.* Iowa State University Press, 1959.
Gruber, H.E., and Barrett, P.H. *Darwin on Man.* Dutton, 1974.

Hamilton, W.D. The Evolution of Altruistic Behavior. *American Naturalist* (1963): 97:354-56.
Harris, S. *The Moral Landscape: How Science Can Determine Human Values.* Free Press, 2010.
Henderson, H., Houston, J., and Hubbard, B.M. *The Power of Yin.* Cosimo Books, 2007.
Henderson, H. *Beyond Globalization: Shaping a Sustainable Global Economy,* Kumarian Press, 1999.
Ho, M.W., and Saunders, P., Eds. *Beyond Neo-Darwinism.* Academic Press, 1984.
Hubbard, B.M. *Conscious Evolution.* New World Library, 1998.
Huxley, J. *Evolution: The Modern Synthesis*. Boston: MIT Press, 2009
Huxley, J. Essays of a Humanist. Harper and Row, 1964

Jantsch, E. *The Self-Organizing Universe*. Pergamon Press, 1980.

Kant, I. *The Critique of Practical Reason.* New York: Macmillan, 1993.
Keltner, D. *Born to Be Good: The Science of a Meaningful Life.* Norton, 2009.
Keltner, D. , and Marsh, J. (Eds.) *The Compassionate Instinct: The Science of Human Goodness.* Norton, 2010.
Keynes, R. *Darwin, His Daughter, and Human Evolution.* Riverhead, 2001.
Krippner, S. *Human Possibilities.* Doubleday Anchor, 1980
Kropotkin, P. *Mutual Aid: A Factor of Evolution.* Porter Sargent, 1950.
Kropotkin, P. *Ethics: Origins and Development.* Dial Press, 1924
Kauffman, S.A. *At Home in the Universe.* Oxford University Press, 1996.
Kung, H., and Kuschel, K.J., (Eds.) *A Global Ethic: The Declaration of the Parliament of the World's Religions.* SCM Press, 1993
Kung, H. *A Global Ethic for Global Politics and Economics.* SCM Press, 1997.

Lakoff, G. *Moral Politics.* University of Chicago Press, 2002.
Laszlo, E. *Evolution: The General Theory.* Hampton Press, 1996.
Laszlo, E. *The Whispering Pond.* Element Books, 1996.
Laszlo, E. *The Connectivity Hypothesis: The Foundations of an Integral Science of Quantum, Cosmos, Life, and Consciousness.* Albany, NY: SUNY Press, 2003.
Laszlo, E. *Science and the Akashic Field: An Integral Theory of Everything.* Inner Traditions, 2007.
Lerner, M. *The Politics of Meaning.* Perseus Books, 1997.
Lerner, M. *The Left Hand of God.* Harper One, 2007.
Lewontin, R.C., Rose, S. and Kamin, L. *Not in Our Genes.* Pantheon, 1984.
Lorenz, E. Irregularity: A Fundamental Property of the Atmosphere. *Tellus*, 36A, pp.98-110.
Lovelock, J. *Gaia: A New Look at Life on Earth.* Oxford University Press, 2000.
Loye, D. *The Healing of a Nation.* Norton, 1971.
Loye, D., and Rokeach, M. Ideology, Belief Systems, Values, and Attitudes. In *The International Encyclopedia of Neurology, Psychiatry, Psychoanalysis and Psychology.* Van Nostrand, 1976.
Loye, D. *The Leadership Passion: A Psychology of Ideology.* Jossey-Bass, 1977.
Loye, D., and Eisler, R. Chaos and Transformation: The Implications of Natural Scientific Nonequilibrium Theory for Social Science and Society. *Behavioral Science* (1987): 32, 1, pp.53-65.
Loye, D. *An Arrow Through Chaos.* Park Street Press, 2000.
Loye, D. The Moral Brain. *Brain and Mind 3* (2002): 133-150.
Loye, D., (Ed.) *The Great Adventure: Toward a Fully Human Theory of Evolution.* SUNY Press, 2004.
Loye, D. *Measuring Evolution.* Benjamin Franklin Press, 2007.
Loye, D. *Darwin's Lost Theory, 2nd Edition.* Benjamin Franklin Press, 2010.

Mackintosh, S. *Evolution's Purpose: An Integral Interpretation of the Scientific Story of Our Origins*. Select Books, 2012.
MacLean, P. *The Triune Brain in Evolution: Role in Paleocerebral Functions*. Plenum Press, 1990.
Maslow, A. *The Farther Reaches of Human Nature*. Viking, 1971.
Mazur, S. *The Altenberg 16: An Exposé of the Evolution Industry*. North Atlantic Books, 2010
Mazur, S. *Royal Society: The Public Evolution Summit*. Caswell Books, 2016.
Min, J. , Ed., *The Chalice and the Blade in Chinese Culture*. Chinese SocialSciences Publishing House, 1995.
Monod, J. *Chance and Necessity*. Vintage, 1971
Monod, J. *Beyond Chance and Necessity*, John Lewis, Ed. Teilhard Centre for the Future, 1974.
Montagu, A. *The Direction of Human Development*. Hawthorn, 1970.
Montagu, A. *Touching*. Harper Collins, 1979.

Narvaez, D. *Neurobiology and the Development of Human Morality: Evolution, Culture, and Wisdom*. Norton, 2014.
Narvaez, D. .*Embodied Morality: Protectionism, Engagement and Imagination*. Palgrave Macmillan, 2016
Noddings, N. *Critical Lessons: What our Schools Should Teach*. Cambridge University Press, 2006.

Pribram, K. *Brain and Perception*. Erlebaum, 1991.
Pribram, K. On Brain, Conscious Experience, and Human Agency. In Loye, D., Ed., *The Evolutionary Outrider*. Praeger, 1998.
Pribram, K. *The Form Within: My Point of View*. Prospecta, 2013.
Prigogine, I. and Stengers, I. *Order Out of Chaos*. Bantam, 1984.

Reich, W. *Function of the Orgasm*. Farrar, Strauss, and Giroux, 1973.
Richards, Evelleen. *Darwin and the Making of Sexual Selection*. University of Chicago Press, 2017.
Richards, R.J. *Darwin and the Emergence of Evolutionary Theories of Mind and Behavior*. University of Chicago Press, 1987.
Richards, R.J. *The Meaning of Evolution*. University of Chicago Press, 1992.
Richards, Ruth. (Ed.) *Everyday Creativity and New Views of Human Nature*. American Psychological Association, 2007.
Richerson, P., and Boyd, R. *Not by Genes Alone*. University of Chicago Press, 2004..
Romanes, G. *Darwin and After Darwin*. Longmans Green, 1893.
Rose, H. and Rose, S. (Eds). *Alas, Poor Darwin: Arguments Against Evolutionary Psychology*. Harmony, 2000.

Salthe, S. *Development and Evolution*. MIT Press, 1996.
Sober, E., and Wilson, D.S. *Unto Others: The Evolution and Psychology of Unselfish Behavior*. Harvard University Press,1998.
Sorokin, P. *The Ways and Power of Love: Types, Factors, and Techniques of Moral Transformation*. Templeton Foundation Press, 1954.
Sternberg, R.J., and Barnes, M.L. *The Psychology of Love*. Yale University Press, 1989

Trivers, R. The evolution of reciprocal altruism. *Quarterly Review of Biology*. 46 (1): 35–57. 1971.

Varela, F. *Ethical Know-How*. Stanford University Press, 1999.

Weber, M. The Social Psychology of the World's Religions. In Parsons, T, Shils, E, Naegles, K, and Pitts, J, Eds. *Theories of Society*. The Free Press, 1961.
Wilber, K. *The Atman Project*. Quest Books, 1980.
Wilber, K. *Grace and Grit*, 1991.
Wilber, K. *A Brief History of Everything*. Shambhala, 1996.
Wilber, K *The Marriage of Sense and Soul: Integrating Science and Religion*. Harmony, 1998.
Wilber, K. *A Theory of Everything: An Integral Vision for Business, Politics, Science and Spirituality*. Shambhala, 2001.
Wilber, K. *The Religion of Tomorrow*. Shambhala, 2017.
Wilber, K. *Trump and a Post-Truth World*. Shambhala, 2017.
Wilson, D.S. and Sober, E. Reintroducing Group Selection to the Human Behavioral Sciences. *Behavioral and Brain Sciences* (1994): 17, 585-608.
Wilson, D.S. *Darwin's Cathedral: Religion as a Multi-level Adaptation*. University of Chigago Press, 2002.
Wilson, D.S. *Evolution for Everyone*. Delta, 2007.
Wilson, D.S. *Does Altruism Exist?: Culture, Genes, and the Welfare of Others*. Yale University Press, 2015.
Wilson, E.O. *Sociobiology: The New Synthesis*. Harvard University Press, 1975.
Wilson, E.O. *Human Nature*. Harvard University Press,1978.
Wilson, E.O. *The Diversity of Life*. Harvard University Press, 1992.
Wright, R. *The Moral Animal*. Random House,1994.

Zimmerman, M. Combating the fifth wave of creationism: religious leaders and scientists working together. *Theology and Science,* 8:2, 211-222, 2010.
Zimmerman, M., and Loye, D. Science and religion: a new alliance to combat the new wave of creationism. *World Futures: The Journal of General Evolution* 67: 1-10. 2011.

REGARDING ERRATA

As happened to Darwin and countless others, rushing first editions into print generally leaves errors in its wake. Identified as "errata," they become corrections for later printings.

For those we've left behind in this book I apologize, but do want to point to this great advantage in being an early reader. This gives you an undeniably sacred first edition, which can immensely soar in value if the book becomes of lasting importance.

Please join all those I write of in this book in talking, writing, teaching, and in other ways working toward this vital end.

Lightning Source UK Ltd.
Milton Keynes UK
UKHW02f2047260418
321724UK00016B/359/P

9 780692 984024